P9-CSE-208

SIGNATURES OF LIFE

SIGNATURES OF LIFE
SCIENCE SEARCHES THE UNIVERSE

Edward Ashpole

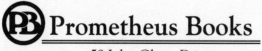
Prometheus Books
59 John Glenn Drive
Amherst, New York 14228–2119

Published 2013 by Prometheus Books

Signatures of Life: Science Searches the Universe. Copyright © 2013 by Edward Ashpole. All rights reserved. No part of this publication may be reproduced, stored in a retrieval system, or transmitted in any form or by any means, digital, electronic, mechanical, photocopying, recording, or otherwise, or conveyed via the Internet or a website without prior written permission of the publisher, except in the case of brief quotations embodied in critical articles and reviews.

Trademarks: In an effort to acknowledge trademarked names of products mentioned in this work, we have placed ® or ™ after the product name in the first instance of its use in each chapter. Subsequent mentions of the name within a given chapter appear without the symbol.

Cover design by Nicole Sommer-Lecht
Cover image © 2012 Media Bakery

Inquiries should be addressed to
Prometheus Books
59 John Glenn Drive
Amherst, New York 14228-2119
VOICE: 716-691-0133
FAX: 716-691-0137
WWW.PROMETHEUSBOOKS.COM

17 16 15 14 13   5 4 3 2 1

Library of Congress Cataloging-in-Publication Data

Ashpole, Edward.
    Signatures of life : science searches the universe / by Edward Ashpole.
        pages cm
    Includes bibliographical references and index.
    ISBN 978-1-61614-668-9 (cloth)
    ISBN 978-1-61614-669-6 (ebook)
    1. Life on other planets. 2. Exobiology. I. Title.

QB54.A8325 2013
576.8'39—dc23

                                                                    2012044572

Printed in the United States of America

To all those Homo sapiens *who look to science to answer fundamental questions like "How do we happen to be here?" and "Is life universal?" And also to those who would like to know whether we really are the star turn after ten billion years of galactic history or just the latest intelligent species to build a planetary civilization that might in time discover what it's all about.*

CONTENTS

ACKNOWLEDGMENTS   11

Chapter 1. LOOKING FOR UNIVERSAL LIFE   13

Chapter 2. BIG QUESTIONS   17
    No Right Frequency   20
    A Very Strange Story   24
    The Unity of SETI   31

Chapter 3. FACING THE FACTS   35
    New Developments   36
    The BBC Showed the Way   37
    A Limited Principle   40
    A Worrying Situation   45
    The Astronautical Theory   46
    What's Out There?   47
    Earth without Humans   49
    Propulsion Chauvinism   50
    Look Out for Robots   52
    Robots Increase the Colonization Factor   53
    Better Robots by Natural Selection?   55
    Only One Certainty   57
    That Something Special   59
    How Many?   59
    Why Now?   61

Chapter 4. THE LIFE OF ALIENS   65

   Electronic Brains   69
   No Food for Aliens   73
   Codes for Life   74
   Making Exotic Life   77
   Here by Chance   81
   The Trouble with Humanoids   82
   Bogus Bodies   86
   The Roswell Legend   88
   Paths to Technology   90
   Unknown Life-Forms   93
   Theologians, Evolution, and the Multiverse   97
   The Fundamental Constants   98
   Life's Machines   102
   The Great Mystery   104
   The Timescale for SETI   106
   One Way to Life?   107
   No Dice   108
   Big Brains and the Robots   109
   Throwing Stones Develops the Brain   111
   Exotic Life   113

Chapter 5. WHERE ARE THEY?   115

   Tea with the Aliens   117
   The Broadcasting Giveaway   118
   The Ozone Giveaway   119
   Finding More Distant Worlds   120
   That Wide Window of Opportunity   122
   Timeless Targets   125
   What to Look For?   128
   The Guessing Game   129
   Ceilings   133
   A Ceiling for Robots   136
   Space Arks   137

Any Recent Visits?   138
Back Engineering   141

Chapter 6. TESTING TIME   143
Alien Archaeology   146
The Dreaded Min-Mins   149
Repeatable Results   150
Where Saucers Have Landed   151
Dutton's Theory   153
Convenient Frequencies   156
Microwaves and Close Encounters   158
Life Markers among the Stars   161
Planets for Life   162
SETI at Home   167
Project Phoenix   168
SETI Sees the Light   169
Photonstar   170
The Lesson from Twisted Light   172
Answers on the Moon and Mars?   174
The Lunar Lights   176
Artifacts on Earth?   180
Landings on Ice   181
Photographic Evidence   183
Mysterious Lights   185
Abductions and Artifacts   188
Bring in the FBI   196
Going around in Circles   197
Who's Carving up Our Cattle?   201
The Science of Options   202

Chapter 7. MYTHOLOGY AND REALITY   205
Good-bye to Classic Cases   208
The Wrong Questions   209
The Fantasy Sect   211

10   CONTENTS

Welcome to All Aliens   212
Aliens Near and Far   214
Inexplicable Technologies   215
Books beyond the Fringe   216
One Way Forward   218

NOTES AND REFERENCES   221

BIBLIOGRPHY   225

RECOMMENDED WEBSITES   227

Acknowledgments

Anyone reading this book will soon see how many different scientific disciplines come together to support the hypothesis that life is a universal phenomenon. So I've needed lots of help and information from lots of clever people over many years. There have been scientists in mainstream science who—knowingly and unknowingly—have provided relevant data, and those who are openly active in what is currently seen as a fringe area for research. Without these fellow travelers, I would never have written this book, and their respective contributions toward answering the major question about life will be evident in the following pages.

CHAPTER 1

LOOKING FOR
UNIVERSAL LIFE

So what is the status of life in the universe? This book examines this question and looks at what science is telling us about how we might discover the different signatures of life and confirm the major hypothesis in human history: *that life and intelligence are universal phenomena.* I like to call this the Grand Hypothesis because they don't come any grander.

Life is the greatest mystery in the universe. The chemical systems and molecular structures that keep every living thing alive are almost beyond comprehension in their complexity, yet they work perfectly. So we have to ask how it is that the universal physics and chemistry has not only created the galaxies, stars, and planets of the physical universe but also formed the basic molecular units that evolution has been able to put together to create complex life. Why, at a molecular level, should everything fit together and work? It's not the outward forms of life that most perplex us, though these are amazing in their infinite variety. It is what continually goes on within every living organism. Each one of us is composed of many billions of cells, and each cell is a factory of marvelous complexity, producing everything needed for life— from proteins to energy. So we have to ask ourselves if all this is the same or similar on other worlds—or substantially different. The range of possibilities must be limited by the universal physics and chemistry and what the process of evolution can do with this. But there's the rub. We don't know what evolution may be able to do. Only the detection of alien life may begin to answer our questions.

Astronomers tell us that the oldest sunlike stars are twice the age of the sun, which means that Earthlike planets orbiting such stars could have supported life a few billion years before its origin on our planet. However, as we know, life can be highly successful without intelligence, and many intelligent species can exist without one having the ability to create technologies. Here we stand alone in this respect, but what about all those other planetary systems? Judging from what has happened here and from the ages of the oldest sunlike stars, technologically intelligent creatures, our counterparts, could sometimes have evolved—and at any time from a few billion years ago. That's an immense period of time during which world civilizations could have developed advanced space programs and explored other planetary systems—at least by robotically controlled craft. So evidence of alien technology could be within the solar system. It seems more likely that intelligent signals are being broadcast from other planetary systems, signals that could be received in our time with our current level of technology. However, evidence of alien broadcasts or alien technology within the solar system would be clear signatures of life and intelligence in the universe.

The fact is that no one knows what the situation is out there or what we might be looking for, but we do have mainstream science to guide us, plus the scientific method to follow. That should be enough for the scientific investigation of ideas and possible relevant data. That is the position taken in this book: that ideas and data must lead to testable hypotheses. Otherwise we are not engaged in science. Testable hypotheses have to be based on what seems possible and on what we already know. A good example of this is the hypothesis that started the scientific discipline of SETI (search for extraterrestrial intelligence). It was published in the science journal *Nature* in September 1959, and its authors, Philip Morrison and Giuseppe Cocconi, two physics professors, suggested that the frequency of neutral hydrogen at 1420 MHz would be chosen by aliens who wanted to communicate with their neighbors. This speculation was based on the fact that hydrogen is more abundant than all other elements put together, and that it radiates this frequency from most of the celestial objects studied by our radio astronomers. Therefore, astronomers everywhere, in whatever life-form, might spot any artificial signals the frequency carried. It was such a persuasive hypothesis that, for the past fifty years, multimillion-dollar programs of radio astronomy have searched for signs of intelligence on that frequency and others near it. A little later, laser technology was developed to a stage where it was suggested that the aliens might prefer to communicate by very brief but intense flashes from lasers rather than by old-fashioned radio. This brought optical astronomers into SETI—and still does.

So both groups of astronomers are looking for evidence on the assumption that, given suitable physical and chemical conditions, life will form on planets and their moons, and that sometimes the equivalent of astronomers will evolve who wish to contact their cosmic neighbors. This might happen in only a tiny proportion of planetary systems, but with a few billion stars like the sun in our galaxy, there could be plenty of planets for the evolution of our fellow travelers in space and time.

However, the searches for radio or laser signals from these other worlds now seem unlikely to succeed for two reasons—and not because our counterparts have never existed. The probable reasons for the failure to find signals to confirm the universal nature of life are the vast expanses of time involved in the history of life on any suitable planet or moon before high intelligence can evolve, and also the speed with which technology develops once such high intelligence exists. These are obstacles to avoid in testing the Grand Hypothesis. And we will see later in this book how this can be done—and is being done. In assessing the justification for "testing," we must keep in mind that world civilizations that might be detectable would be far older than ours and supported by technologies beyond our power to imagine—because we are not yet familiar with the science that would support them. Also, we know from the fossil record and geology that watchful aliens might have detected the Earth as an inhabited planet from its spectral lines at any time during the past 350 million years, and probably for a much longer period. We know that those lines would have been radiating from the Earth during this immense period of time—and they still do—making our planet a target of interest. Consequently, the arrival of evidence such as probes from other worlds during the past 350 million years is not impossible. In fact, it looks a considerably better bet than the arrival of detectable alien broadcasts in our time. Many planetary civilizations could have evolved and become extinct during that period.

Nevertheless, we owe a lot to the scientists in astronomical SETI who still scan the galaxy for signals. They may not have found any aliens during the past fifty years, but they have made us think hard about the status of life in the universe and about the possibilities that other world civilizations exist or have existed in the past. Yet we still don't know if we are a one-off miracle or just a tiny sample of a spontaneous phenomenon that flourishes throughout the universe. And although we can never know if we are a one-off miracle, we can use science and technology to check if life and intelligence are inevitable products of the way our universe works.

But the territory to be explored is different from anything science has tackled before. Consequently the approaches to it are different—and more diverse. In the history of science there can have been few subjects open to attack from so many different angles, though you wouldn't think so from what is being published in the science journals. A few of these angles may seem too acute for comfort, but when the objective is to answer the most important question about life, we have to consider everything that might be relevant. Everyone is on the fringe in this line of research, supported more by speculation about what the relevant science may indicate than hard data. So it's essential to distinguish the "rational fringe" from the "lunatic fringe" and to stick closely to the tried and tested ways of science. It's the results from testing hypotheses that matter. Nothing else. We have to form hypotheses, where that is possible, and test them with the aim of getting repeatable results. But while we're waiting for someone to shout "Eureka," you can review the relevant science and the different projects being run by good scientists to test the Grand Hypothesis.

CHAPTER 2

BIG QUESTIONS

Although we frequently hear questions like "Who are we?" and "Where do we come from?" science already provides reasonable answers—perhaps the best we can have at present. But as we've seen, the big question science has not answered is "What is the status of life and intelligence in the universe?" And since we're the only species on Earth capable of asking that question, perhaps we should try to find the answer.

The phenomenon of life is the product of such physical and chemical complexity that it's not yet understood, despite the revolution in biology of the past fifty years. The mechanisms that keep us going twenty-four hours a day are vastly greater both in number and complexity than those that keep our many technologies working. Evolution has produced biological systems that have been self-sustaining and self-perpetuating on Earth throughout the past three and a half billion years. And everything we have and know, our whole civilization, depends on these amazing organic systems working perfectly to maintain us and all other life on Earth—the entire biosphere.

But life and evolution have only been able to produce our biosphere because the physics and chemistry of the universe are precisely as they are. And we know that the same physics and chemistry will be available everywhere, ready to produce life wherever conditions allow. The information we have points to the probability that the origin of life is inherent in the nature of physics and chemistry. If that is so, then life must be universal because physics and chemistry are universal. We rely on this assumption being a fact in the study of even the most distant parts of the universe, so what happened on Earth some four billion years ago, when the physical conditions were right for the origin of life, would happen anywhere in the universe where the same conditions prevailed.

We may also wonder at how robust life has been on Earth during the past four billion years. Its capacity, once established to survive catastrophe after catastrophe on our planet, makes it look like a universal phenomenon. Most of us may already believe that this is so—that life and intelligence are universal—but belief alone is not enough. Science isn't in the "belief business." We need confirmation. And this is what this book is about: finding confirmation. Many scientists in very different lines of research will be attempting to find this in the coming decades. Astronomers in the discipline of SETI (search for extraterrestrial intelligence) have already been trying to find alien intelligence for decades by searching for broadcasts from other planetary systems. Their research rests on the hopeful assumption that our cosmic neighbors are born communicators who will use radio technology for an eternity to try to contact their lesser counterparts in space and time. These astronomers have numbers on their side—there are a few billion stars like the sun in our galaxy alone—but there's a problem with this approach. Astronomers throughout the galaxy, whether they have limbs or tentacles, would have to be present and ready with the right equipment to receive the transmitted signals (whatever they might be) when they arrive—which, as we will see, could have been at any time during the past four billion years. That's a lot to ask, because astronomers of the humanoid kind have been rather slow to evolve on this planet.

It has taken 3.5 billion years to progress from bacteria to the life-forms we have today. And, according to the ages of the oldest sunlike stars (about twice the age of the sun), the first broadcasters in the galaxy could have been in business by about four billion years ago. Consequently, an amazingly long series of communicating civilizations would be needed for signals to be arriving here today. The problem is that the cosmic timescales of planetary formation and organic evolution are vast and weigh heavily against interplanetary communications by radio or lasers. World civilizations may quickly pass through our level of science and technology millions of years apart, which would make synchronizing speed-of-light radio communications impossibly difficult. But it's not only this "speed-of-light" limitation that would have discouraged budding interstellar communicators during the past four billion years. It's also the forward march of science that could put every world civilization at a different technological level. No one would deny that we will develop better systems of communication in our future, so it's a reasonable conclusion that if "they" are out there, "they" will have already done so. This means that astronomers who search for cosmic messages face a "missing technology" problem as well as a "speed-of-light" problem. It means that if advanced aliens are in reality transmitting messages to other worlds, then we won't yet have the equipment to receive those messages.

What persuasively demonstrates this point is a development at the University of Glasgow, where light and microwaves are being twisted to carry much more data in a highly focused beam.[1] That development will advance communications on Earth—and it's taking place only fifty years after astronomers began to search for alien signals in untwisted microwaves and laser light. So what communication systems might there be far beyond "twisted light," say, in five hundred or five thousand or five million years? These are brief periods in the timescales involved in planetary and biological evolution and probably in the development of civilizations. It's therefore obvious that you can't guess what systems of communication other worlds will be using, so searching for messages on specific frequencies with present-day technologies, as SETI astronomers have been doing for decades, is unlikely to be successful.

It's a catch-22 situation. If the aliens are broadcasting, we don't have the technology to detect their broadcasts. If they're not broadcasting, this strategy in searching for them is wrong, and we should broaden our research. We need to see the very big difference between the possibility of detecting physical evidence of alien intelligence, which might be within the solar system after millions of years, and detecting intelligent signals from the stars. The early expectation, held by many astronomers, that messages from super civilizations will eventually be detected from other planetary systems is beginning to look increasingly unrealistic. We can't help remembering that eminent scientists of past generations have been equally unrealistic, although not at the time they made their suggestions. For example, in 1941, when the Germans were bombing London, Sir James Jeans suggested that we might use our searchlights to flash prime numbers to show mathematically minded Martians that intelligent life existed on Earth, which could have been highly questionable at the time.

NO RIGHT FREQUENCY

The commitment of SETI astronomers to radio technology has meant that they have thought a lot about the right frequencies to scan for intelligent signals. But there can be no "right frequency" to connect us to aliens who are using communication technologies that we can't yet imagine. So there's probably no right frequency for interstellar communications. We've just seen that only fifty years after the first radio searches for extraterrestrial broadcasts, we are developing a radical new way of transmitting radio waves and microwaves that current receivers couldn't pick up. And we are sure to develop more and more advanced systems, as other civilizations would have done in their time. Thus the incompatibility of communication systems stands between us and other worlds, given that, in the first place, interstellar broadcasting with the speed-of-light limitation is a worthwhile activity.

Nevertheless (as we'll see later), SETI astronomers could be essential in the search for evidence of alien life and technological intelligence. We will see that much information about universal life may come from the stars. Astronomers may never receive messages of goodwill and wisdom from distant worlds, but they are admirably equipped to detect other forms of evidence, which could be obtained without worrying about O'Neill, the irksome "missing technology" problem." They might detect the presence of alien probes, or the spectral evidence (at infrared frequencies) coming from suspected "Dyson spheres" (spheres of space colonies named for physicist Freeman Dyson), or signs of similar activities (past or present) from some of the newly discovered planetary systems. These planetary systems become increasingly interesting as more are discovered. One example is the Gliese system, just twenty light-years away. It has four known planets (16, 7, 5 and 1.9 times the mass of Earth) orbiting a star somewhat smaller and cooler than the sun. The smallest planet orbits within the so-called habitable zone of its sun, where water could exist in a liquid state.

Astronomers have already used both space telescopes and ground-based telescopes to search for Dyson-O'Neill spheres, and such searches will continue, especially if biomarkers are detected in other planetary systems. Although Dyson-O'Neill spheres exist only in theory, they look like a natural development where the evolution of technological intelligence has been so successful that it could no longer support its technological activities and its expanding population on its home planet. The need for more living space, as well as the dangers of an overheating planet and the prospect of unlimited free energy from its parent sun may force some civilizations to build and maintain a multitude of space colonies in orbit around the star of a planetary system, so forming a kind of disconnected shell of space colonies. The late physicist Gerard O'Neill, in his work on space colonies, showed more than anyone how such spheres could become an inevitable development over a long period of time.[2] Thus the detection of biomarkers and techno-markers could provide provisional answers to the question that biologists—and everyone else— would like to answer. Are the origin of life and the evolution of technological intelligence universal?

But for the detection of intelligent signals, the "missing technology" problem looks like a permanent barrier. Nevertheless, SETI radio astronomers continue to scan a narrow band of microwave radiation that, several decades ago—when the subject was new and naïveté flourished—some really good scientists thought was the best bet for alien broadcasts. Since then the equipment and technical know-how have advanced beyond anything conceivable when the first searches began, while the basic assumptions about communications across interstellar space have remained largely unchanged. Likewise, astronomers in optical SETI are still looking for messages—but messages encoded in nano-flashes of light.

SETI astronomers have consistently denied the possibility that our cosmic neighbors with the most advanced technologies may have preferred direct exploration to broadcasting. Not even robotic probes are allowed to visit us. It's natural that scientists in astronomical SETI don't like this idea. Their thinking goes like this: "We see no feasible way of crossing the light-years to other planetary systems, so it can't be done. Other intelligent species, even if they had interstellar transport for the odd trip, would think it easier and cheaper and safer to communicate by radio or lasers."

This "belief" that the aliens will want to communicate with anyone bright enough to invent radio is still firmly held. The hope still lingers that the *Encyclopedia Galactica* is being beamed across the galaxy for anyone who can understand it and use it. Of course, if we detected a radio message from the aliens, we'd know they existed. But that's about all the information we might get. To anticipate a free supply of new knowledge from distant worlds conveniently beamed to us in a way completely compatible with our own current astronomical equipment is to expect too much. When we review our present advancing technologies, such as artificial intelligence and space probes, the possibility that our fellow travelers in time and space may already have artifacts in the solar system seems somewhat greater than that they have been broadcasting evidence of their existence throughout major periods in the history of our galaxy, and doing so with our level of radio and laser technologies.

Interstellar travel may only look impossible from where we stand. Go back in time to 1492 and meet Christopher Columbus and his sailors, who have just arrived in the New World after a hazardous and rather stressful two months crossing the Atlantic. You then tell Columbus that in five hundred years, thousands of people every day will cross the Atlantic in seven hours—in flying machines. In 1492, that would have looked just about as impossible as interstellar travel does today, though if you replace flesh-and-blood astronauts with highly advanced robots, it begins to look possible. Anyway, if interstellar spaceflight is physically possible, then the aliens have had plenty of time to develop it. As we shall see: anything up to four billion years. So the scenario that our counterparts have been traveling from planetary system to planetary system for a very long time is not unreasonable. We will see later how "they" could have discovered the presence of the Earth at any time during the past few hundred million years. The science that attests to this immense window of opportunity is readily available. This, of course, doesn't mean that "they" have come. They may have been too far away or never have existed. It only means that scientists should be ready to consider the relevant science and to test the possibility that they have come—because there is vague and untested data that they have.

Here we meet a persistent problem. One facet of this local approach to SETI can involve a source of information associated with the tallest stories around. I'm referring to the UFO phenomena, a subject understandably anathema to astronomers searching for broadcasts from the stars. But it's more than that. Anyone with their neural circuitry in reasonable condition can see that much of what is published about UFOs is unscientific and unreliable—even outright baloney. There are reasonable websites offering data that could interest the science community, but there are others that are either the work of hoaxers or people suffering from UFO psychosis. Some even describe the physical and psychological characteristics of a whole range of different alien species that just happen to be visiting us in our time from different parts of the galaxy. No one has any evidence—not one alien cell or a scrap of tissue to put under a microscope—yet the people who put these pages on the web know all about the anatomy and genetics of these visiting aliens, such as the reptiloids, amphiboids, insectivoids and various other "oids."

So are we being hoaxed or have some UFO lemmings gone right over the cliff? I think most of these web pages are hoaxes because the lunacy of the content does not match the competent level of writing and presentation. One would think that the major UFO associations would stamp heavily on this kind of thing. But no, they still tolerate reports from anyone who claims to talk with the aliens—conveniently by telepathy—about issues that are totally familiar to all of us. The aliens never have anything new to tell us after crossing interstellar space—things we don't know already.

The UFO associations do pay lip service to science, but they don't always insist on acceptable evidence for what is being claimed, which is no less than confirmation of the major hypothesis in science. This has not exactly encouraged scientists to take a serious look at the UFO phenomena, though what has been reported is sometimes strangely consistent with our scientific knowledge, which indicates that an alien presence in the solar system is far from impossible. And some scientists, through investigating the UFO phenomena, have discovered surprising new plasma and atmospheric physics. Yet the question remains: could there be an alien reality hidden in the massive UFO mythology? It could be a mistake to ignore all aspects of the UFO phenomena, though we shouldn't think that those alien humanoids described on the web have anything to do with the subject.

The basic point to remember is that our galaxy is more than twelve billion years old, and that the first sunlike stars and planetary systems would have existed by about nine billion years ago. That means that planets like the Earth that formed nine billion years ago could have supported technological civilizations some four to five billion years later—at about the time when the Earth was formed. Consequently, the solar system has been open to visitors for a very long time, after things settled down. Most scientists would accept this point, given that interstellar space is crossable, at least by advanced artificial intelligence (robots). However, those scientists who currently search for evidence based on this provisional assumption are barely tolerated, and those who take an active interest in the UFO phenomena put themselves on the brink of professional suicide. Yet mainstream science and what we know of the way technology advances support the possibility—only the possibility—that a very small proportion of UFO reports are describing alien spacecraft or space probes. But although it's no more than a possibility, it's something that justifies the attention of science, which demands irrefutable proof before anything is accepted.

A VERY STRANGE STORY

On May 15, 2006, the British government, under the Freedom of Information Act, released its secret scientific study of the UFO phenomena. Its code name is Project Condign and it was produced for the Ministry of Defense. It took four years (1996 to 2000) to complete the four volumes and 460 pages. It deals with natural and human-made explanations for UFO reports and finds plenty of them: clouds of all kinds, aircraft condensation trails, seismic activity, lightning, meteors, covert aircraft testing, hang gliding, plasma balls, birds—even clusters of large moths. This is nothing new. Those familiar with the subject have always accepted that such things must account for a vast number of UFO reports—though the moths are new to me.

Plasma-type phenomena in the atmosphere—which are good candidates for UFO reports—get special attention without much in the way of scientific explanations. This is not surprising. When I last wrote on this subject, I talked with several plasma physicists at different universities in Britain. At the time, none of them knew how plasma balls formed, and none thought these could survive for long in the atmosphere. However, major media channels in Britain and elsewhere went along with what the study had to say about plasmas and everything else, presumably because it was a government report, though on other issues the media is quick to query such documents. For example, one BBC journalist reported, as if saying something significant: "A confidential Ministry of Defence report on Unidentified Flying Objects has concluded that there is no proof of alien life forms." Similar statements came from elsewhere. The point is that this part of the ministry's conclusion is totally obvious. If there were proof, the Grand Hypothesis would be confirmed and the greatest breakthrough in the history of science and humankind would have been made. The reason for the present various lines of scientific research in SETI is to find proof, if proof exists to be found. Paper studies like Project Condign are not going to prove anything. You have to do scientific research to make scientific discoveries. And you can't claim an armchair conclusion as a scientific finding.

However, Project Condign does comment on all kinds of poorly understood phenomena in the atmosphere, some of which must certainly be responsible for numerous UFO reports, including some aircrew reports, and that should make us more careful about what we accept as possibly credible. But natural plasmas and other natural phenomena in the atmosphere could hardly perform all the amazing tricks that aircrews have reported over many years. Plasma balls are rather delicate creatures, unsuited to hurtling through the atmosphere at hundreds of miles an hour to outpace aircraft while maintaining their integrity. Plasmas are also rather scarce in the atmosphere in daylight when the weather is good and the skies clear. And it's difficult to explain how aircrews could see a ball of plasma as a metallic-looking structured craft. So it's a problem that 460 pages of documentation from the Ministry of Defense do not solve.

At first glance, therefore, Project Condign seemed puzzling. Of course, like all who have tried before, the study couldn't explain all aspects of the UFO phenomena. That's understandable, but much relevant material is missing. I envisaged numerous high-powered scientists being called to the ministry to provide the fruits of their expertise and knowledge. But it wasn't like that at all, according to Dr. David Clarke of Sheffield Hallam University (who used the Freedom of Information Act to free the document for the public). Only one person was involved. The unknown author was commissioned to prepare the study in secret and alone—no consultations with scientists and others allowed. He couldn't do any research himself. He could only read what others had done—and no one had done enough . So how could he help but miss chunks of the subject? For example, there's no indication that he knew about the catalogs of pilot and aircrew reports that are a major source of data. Just as well, really, because some of that data would not have fit the explanations offered. The data exists in several catalogs, but there are three catalogs that classify more than three thousand cases in which aircrews have encountered strange aerial phenomena. The main catalogs exist through the work of NASA scientist Dr. Richard Haines, now retired, and Dominique Weinstein, technical adviser to the National Aviation Reporting Center on Anomalous Phenomena, who began his work with the French Space Agency.

So how did it come about that one man was asked to prepare a four-volume report on such a wide-ranging scientific subject for the Ministry of Defense? We can only guess. Picture the scene in 1996: two civil service mandarins are talking at the ministry.

A. I'm worried. The prime minister keeps asking about all those UFO stories he reads about in the papers. He's taking it all very seriously. I think he might become a ufologist.

B. Well, the minister would never stand for that.

A. Quite. We can't have ufologists in the government. That would never do.

B. Precisely, so let's get someone to review UFOs for the minister, a proper scientific chap. Someone sound who can do the job alone and secretly. We don't want any journalists getting wind of this and thinking we're worried about aliens threatening Britain.

A. Or the minister thinking those stories in the newspapers might be true.

B. Yes, that would never do. But people can believe anything these days. Did you know that some American astronomers have been listening for radio broadcasts from other worlds for fifty years?

A. What, for fifty years? They must be getting a bit tired by now.

B. Well, yes, I suppose they are. But they're doing a good job, really. If the aliens aren't out there broadcasting, they could be here. That's the problem, you see. And we don't want that, do we—people thinking that aliens are here in UFOs? So it's best to look for them broadcasting as far away as possible. People won't worry then.

A. I see. So the scientists say the aliens are so far away that they can only contact us by radio, then everyone will think they're out there and not here.

B. Yes, that's it. The aliens may not be threatening us anyway, but the public might think that they could—and would. That's the danger.

A. Right, so we have to get a really sound chap to squash that idea. Someone who wouldn't get any funny ideas in the course of the work. And we have to keep his work secret until we're sure he's reached the right conclusion.

B. Yes, that's it. Then we can leak that the document exists, and someone is sure to ask for it under the Freedom of Information Act.

A. Excellent. We then let it go, the act is validated, and everyone gets the right idea. So who do we get?

B. What about old Henry at Oxford? He's teaching physics there, or is it physiology?

A. Yes, old Henry could be the right chap. His wife is a vicar, you know. That could keep him from getting strange ideas about alien visitors. The Church wouldn't want anyone looking for aliens. They might have their own religion, and we've got enough religions in this country already.

B. Exactly. So just get old Henry to do a really thorough job on the subject—about four hundred to five hundred pages would do. That should keep the minister quiet for a while.

A. Quite right. We don't want him joining the British UFO Association or any-thing like that. It could bring down the government.

B. And the civil service. People would begin to think we're out of touch or something.

And that's how "old Henry" got a job he couldn't possibly do. To start with, he faced the problem that many have faced before: you cannot explain every UFO case in terms of human-made objects and natural phenomena, even if you understood the relevant natural phenomena. And as no one really understands the relevant natural phenomena, poor old Henry didn't stand a chance. Also, Henry was on his own. He couldn't talk with anyone in case he blew his cover. Four years toiling away with only scientific papers to consult would have been too much for most of us, but he eventually filled 460 pages that looked good enough to anyone not familiar with the subject. Of course, the term *UFO* was a bit of a problem for him, but he replaced it where he could with *UAP* (unidentified aerial phenomena). That was a good move, bearing in mind the baggage of nonsense carried by the old term. However, Henry missed the most important aspect of the subject—or he thought it wise to leave it out. What most supports the need to check the best UFO data is the scientific rationale (the SETI rationale) that has justified the high-powered astronomical searches for ET (extraterrestrials) for the past fifty years. Alien artifacts may or may not exist in the solar system, but the SETI rationale is the only sound scientific reason for thinking that there might be something to discover.

The position should be that when we're testing the major hypothesis in science, we should feel free to use any data that makes scientific sense. But programs of alien-human breeding and telepathic communications with the occupants of flying saucers do not make scientific sense. So we have to be selective and also accept that even reports that seem credible may be misleading. The only way out of this problem—for scientific research on this difficult subject—is to make a sharp distinction between "usable data" and "acceptable data." By this I mean that we are justified in using plausible data (usable data) that cannot be confirmed as true and accurate, providing it is used only to form testable hypotheses; it is then the results from testing that count. Either the results will confirm the reliability of the data or they will not. Later in the book, we'll see how this can work in a conventional scientific way.

Most scientists, as I've discovered over the years, are not familiar with the scientifically interesting aspects of the UFO phenomena—and why should they be? They rank media stories about UFOs as about as reliable as the astrology columns, so they haven't considered the theoretical links between mainstream science and those aspects of the UFO phenomena that, for example, allow for the possible existence of alien probes. This possibility—though it's just a possibility—deserves to be rigorously tested, but so far only one observatory has done so. Eamonn Ansbro's Kingsland Observatory in Ireland is specially equipped with eleven cameras to automatically record any probes in orbit. It is also designed to test Roy Dutton's astronautical theory (more on that shortly). Scientists in the SETV (search for extraterrestrial visitation), which had its origin in the famous Jet Propulsion Laboratory at NASA, are also planning to search for alien probes. But they don't want to get drawn into the UFO mythology. However, if UFOs are real, then they would presumably come from locations in the solar system, perhaps in Earth's orbit. In other words, the UFOs, or the vehicles they came in, would be alien probes of some sort. It's not impossible, so the idea needs testing, but most scientists and all their governing organizations shrink from anything associated with "UFOs."

Too many books and articles have promoted this fantasy-laden term. And the situation is getting worse. There was a time when editors stood between authors with wild imaginations and innocent readers. No longer. There are some good UFO websites, such as UFO Evidence and UFO Skeptic, but others, if not hoaxes, look very suspicious. I actually read that the "gray aliens," the ones with "biomechanical black eye coverings," are upgrading their reptiloid genes with DNA from human abductees. They want to be more like us, although being gray and skinny with a large head and permanent dark glasses may make that difficult.

Then there are the humanoid amphiboids in South America to worry about. One amphiboid told an abductee in passable Spanish that Earth was their original home, but that our distant ancestors drove them into the fifth dimension. This is better than we could do today if the amphiboids became troublesome. I think a particular group of physicists working at CERN (European Organization for Nuclear Research) would like to know about that. They're searching for the fifth dimension in the breakup of nuclear particles and would be rather surprised to find amphiboids lurking there. I don't know how large the amphiboids are, but they'd be rather cramped in the fifth dimension, if the physicists are right. They say the fifth dimension is too small to be detectable in our everyday experience and that we would need the most powerful particle accelerators to find it. Of course, just because physicists say the fifth dimension should exist, we don't believe them until they've confirmed it experimentally and others have checked up on the discovery. But among "true believers" in the UFO jumble, the amphiboids can live in the fifth dimension and no evidence is needed.

Things get worse on some UFO websites where aliens associate with multidimensional poltergeists and even bioplasmic vampires. All this somewhat dampens scientific enthusiasm for UFOs. The possible presence of vampires is not something to keep most scientists awake at night. So it's all a big pain for those who suggest the possibility that alien artifacts might be detectable in the solar system, and that the Grand Hypothesis might be confirmed in this way, rather than by searching for messages from the stars. Of course, human brains are suggestible brains, and some are far more suggestible than others. And this has been a problem for ufology, which has attracted a lot of suggestible brains. Given a thorough soaking in the UFO mythology, with occasional helpings of *Star Trek,* many folk seem primed to see flying saucers whenever there's an opportunity, though this may not apply when aircrews report seeing structured craft at the same time that ground radar stations confirm their reports. We have to lay emphasis on "structured craft" because many sightings from aircrews could be explained as natural phenomena in the atmosphere, like luminous balls of plasma.

One surprising feature of our friend Henry's UFO study is that the electrical charge of aircraft exhausts can attract plasmas from up to ten kilometers away—although he doesn't explain how they can do this or how plasmas can follow aircraft at hundreds of miles an hour without dissipating. One might think that this would happen rather quickly. However, according to Henry, a ball of plasma "is likely to behave as a non-deformable solid sphere." His study goes on to say that Russian aviation authorities advise pilots about this: "If the plasmas appear ahead of the aircraft, while it is impossible to lose the attracted entity by acceleration, it is nevertheless possible to carry out manoeuvres to make the mass fall astern, where it will remain with no further danger to the aircraft."

No one would deny that sources of energy in the atmosphere produce phenomena such as balls of plasma that behave in ways that could be mistaken for strange aerial craft. But could all UFOs reported by aircrews be only natural atmospheric phenomena? Not if many of the reports are accurate. Apart from structured craft, there are so many reports of luminous objects that behave in ways that seem to defy the laws of physics. All this data exists in the catalogs prepared over many years by scientifically minded people, but "old Henry," working on his own for the ministry, seems not to have known they existed—and that the much of the data does not support all his theories.

THE UNITY OF SETI

We can see by now that any testing of the Grand Hypothesis needs a sound scientific base, and this exists in the SETI rationale. The SETI astronomers who have scanned our galaxy for intelligence signals for fifty years have used a rich combination of data from several scientific disciplines to justify their research. But, as we shall see, this SETI rationale better supports the possibility of an alien presence in the solar system than the possibility that messages from distant worlds may be detectable by present-day astronomical techniques. Outside the science community, the extraterrestrial explanation for UFO phenomena has been widely accepted, but this acceptance has been no more than a "belief" that has been embellished beyond reason by a multitude of UFO enthusiasts. It has yet to be tested in ways that major claims in science are tested. Yet no one can avoid the fact that the science that has justified astronomical SETI for the past fifty years also justifies the extraterrestrial hypothesis for some aspects of the UFO phenomena. It doesn't justify belief, but it does justify proper scientific investigations. I would add that if the science that has justified astronomical SETI did not exist, the UFO phenomena would not be interesting—at least not to me. But the synthesis of science assembled by the SETI community during the past fifty years does exist, and for that reason we should include the UFO phenomena with the other sources of data that are available to us to test the Grand Hypothesis.

Of course, doing research on the UFO phenomena is one thing, while the implications of getting a positive result would be something very different. If some UFOs have a physical reality, if intelligence were to be behind just one in a thousand of the reported phenomena, then humanity is faced with a dramatic paradigm shift. But that, at present, is only a possibility. Paradigm shifts don't come easily. They are only the result of irrefutable scientific evidence. So in this book I am setting out the synthesis of science we have to consider that might indicate a forthcoming paradigm shift, and at the diverse research projects that may lead to that shift. That may come about or it may not. But whatever happens in the coming decades this is a subject worth following.

However, even if the most dramatic confirmation of the Grand Hypothesis were to be obtained, there would be little need to worry. For statistical reasons, an extraterrestrial presence could not have just arrived, if it has ever arrived. It would have to have been present for a very long time—probably for millions of years. Yet if this was so, our position in this universe would be distinctly different from what we have assumed, and we would have to cope with a third intellectual revolution. We all know that Copernicus and Galileo changed the view we had of ourselves in relation to the sun and the universe. Darwin and other biologists then showed that human beings evolved from earlier life-forms and are therefore an integral part of all life on Earth. These were two revolutions in our view of ourselves.

The third revolution would come from definite evidence of extraterrestrial life and intelligence. The detection of signals from transmitters light-years away might not worry us. Likewise, the detection of fossilized alien artifacts millions of years old within the solar system could be viewed with some detachment. But knowing that superior beings in flying saucers are keeping an eye on us would be rather different. It would generate more attention and apprehension than Copernicus and Darwin ever did; though, as we'll see later, the background information on this subject shows that we should have no need for concern; however, not everyone would be familiar with this. Anyway, the intellectual implications would be greater than our world has ever known, which is why evidence would have to be irrefutable, something that the web page inventors of the reptiloids and the amphiboids don't worry about.

The cardinal rule of science applies here: the greater the importance of a hypothesis, the greater the burden of proof. And we are dealing here with the major question in science. Many rational and intelligent investigators in ufology seem not to have realized this: that verifiable evidence is needed before you can claim confirmation of the major hypothesis in science. And too many people are ready to accept, without proof, that the aliens are either here or out there. Gullibility has been rife. This is not a crime, of course, but many authors of UFO books would be in jail if it was. One way forward would be to link scientific research on certain aspects of the UFO phenomena with the remarkable technical developments and know-how of astronomical SETI. And this, I think, might be done when we have to look at new ways to test the Grand Hypothesis. Some adventurous astronomers will certainly be needed, but, as we'll see later, they won't be playing the same game they expected to play in the early days of SETI.

CHAPTER 3

FACING THE FACTS

S o what about testing the Grand Hypothesis? Scientific hypotheses don't come any grander, but for many, it will seem too grand, too far removed from practical scientific research, something speculative and remote that will not provide new knowledge. But in our time the hypothesis that *life and intelligence are universal phenomena* is scientifically testable. We may not get any positive results, but if we don't test, we certainly won't. Philosophers and theologians have debated the subject for centuries, but without the necessary science and technology, that's all they could do—debate. In our time, the situation is different: today's science from astronomy to biochemistry enables us to think productively about the subject while new technologies offer ways to test. We know (simply because we're here) that the universal physics and chemistry makes the origin of life possible, and that the phenomenon could be a long-lasting feature on any planet that provides appropriate conditions.

But to confirm that what has happened here has also happened elsewhere, we have to find the signatures of independent life beyond the Earth. Until then, we could be a unique chance event—a miracle. Thus the search for evidence is a quest for the greatest discovery in human history, and some of the brightest scientists, having recognized this fact, have become involved. But in five decades and many research projects, the astronomers have found not a single sign that ET is out there among the stars. So are we alone in our galaxy? Are the nearest civilizations thousands of light-years away, too distant to be detectable? Are they too advanced to bother about new civilizations? Are they sending messages across the galaxy but with technology we won't have for a thousand years? Or are we not looking for detectable evidence in the right

35

way? Obviously, researchers can only proceed on the basis that there is evidence to detect, so we will proceed on that assumption while emphasizing the pressing need to broaden the search.

What makes this subject worthwhile is that it's not just a search for ET. The science that forms the rationale for SETI helps us to think about ourselves in this universe. We have an inborn curiosity about our own position on a rather special planet and how we may relate to the phenomenon of life and the whole universe. Without the relevant science, we couldn't think about such things. So there's an immediate value to this subject of SETI, quite apart from the fascination of the searches in progress and those in the planning stage to find evidence of alien life and our fellow travelers through time and space. But there is this strange inconsistency in the science establishment. It looks favorably on searching for ET among the stars, but not on the possibility that anything alien could be within the solar system, though this is possible, judging from the relevant science and the ways in which our own technologies are developing. The problem has been that the science community quite rightly expects research papers that give evidence to back up claims that anything alien exists in the solar system, be it structures on the Moon and Mars or flying saucers. And the UFO associations have rarely provided such research papers. Rather, they have focused on thousands of reports of strange phenomena and the detailed collection of data. Now, data collection is the basis of research, but that alone is not enough. I have the impression that most ufologists don't appreciate the enormity of the claim they are making when they stress the credibility of witnesses and accept accounts of contacts with aliens as true and proven. If this were so, then life and intelligence would be confirmed as abundant throughout the universe, and the billion-dollar projects to test the truth of this assumption would be no more than supplementary exercises.

NEW DEVELOPMENTS

The "indisputable proof" being sought may come from new scientific research projects that are developing in remarkable and unexpected ways. Since astronomers first began in the 1960s to search for alien broadcasts, their subject has become a highly developed discipline. Even though there may be no alien broadcasts to detect, it is nevertheless recognized as a proper subject for astronomers to pursue. It has its own "section" in the International Astronomical Union, and no one is going to be banished from the science community for being a SETI astronomer. But the number of scientists who run the risk of banishment, who favor a more local and direct approach to testing the Grand Hypothesis, is increasing. Heresy is on the march. Rather than searching for hypothetical broadcasts from our betters in the galaxy, some scientists are searching for evidence nearer home. And there are several ways by which this evidence might be found. Like all rebels, these scientists who pursue local SETI follow an uncertain path, but then everyone in SETI follows an uncertain path.

In a paper in *Nature*, Jack Cohen and Ian Stewart said, "The history of science indicates that any discussion of alien life will be misleading if it is based on the presumption that contemporary science is the ultimate in human understanding."[1] That wise statement should be pinned to the office wall of every SETI astronomer. But we are going to be misled. It's inevitable. And the searches for alien broadcasts using our current radio technology are probably one example of this. That is why we have to focus on phenomena rather than on what we think the aliens may be up to. Besides searching for what today's science tells us might be present, we should be ready to investigate any inexplicable phenomena that just might indicate an extraterrestrial source. However, astronomical SETI is not faced with redundancy. It's going to be essential in the new research projects to test the Grand Hypothesis.

THE BBC SHOWED THE WAY

The BBC in London may not wish to be associated with the search for aliens, but my first contact with this subject was a BBC radio program in the late 1940s that proposed the use of radio to communicate with other worlds. I remember listening to this on our 1930s set, intrigued by the prospect of life on other planets, though at that time other inhabited worlds were about as remote as heaven. The nearest we came to that subject was watching Flash

Gordon's trip to Mars at the local cinema. There was then no conceivable way by which science might discover the real thing. But move on fifteen years, and I was reviewing a book by Sir Bernard Lovell, who was then director of the Radio Astronomy Observatory at Jodrell Bank. Sir Bernard's book dealt with conventional astronomy, but he mentioned the now-famous paper "Searching for Interstellar Communications" by Philip Morrison and Giuseppe Cocconi, two American physics professors. Their paper in the journal *Nature* in September 1959 suggested that the frequency of neutral hydrogen at 1420 MHz (the most common natural frequency of the universe) would be an obvious choice for aliens who wanted to let others know that they weren't alone in the galaxy. Hydrogen is far more abundant than all other elements together and radiates this frequency from most of the celestial objects studied by radio astronomers. It reveals so much information that we might expect astronomers everywhere, of whatever life-form, to observe it and therefore to spot any artificial signals that the frequency carried.

Anyway, it seemed such a good bet for interstellar communications that Morrison and Cocconi asked for searches to be carried out. They wrote to Sir Bernard, who declined to use the facilities at Jodrell Bank. He thought the chance of success was too small. However, it is now part of history that the American astronomer Frank Drake thought differently and searched for three months in 1960, using the eighty-five-foot radio dish at the National Radio Astronomy Observatory at Green Bank in Virginia. He didn't find any alien broadcasts, but he did find vast numbers of people on Earth who were fascinated by his search. From that time on, some of the best brains in science developed a convincing scientific rationale for further astronomical searches. Clearly, if the scientific rationale was correct, then SETI could offer an opportunity to test the hypothesis that life and intelligence are universal, something most people have thought about at some time. And the discovery of an alien broadcast, in addition to being an academic achievement, might provide free advanced courses in sciences and technology from a superior alien world. Anyway, that seemed a reasonable belief at the time.

In the 1960s and 1970s, the world was enthralled by the space programs, so the ideas of SETI fit in well with the interests and expectations of the time, and numerous searches began, mainly in the United States and the Soviet

Union. Today the searches continue in a more advanced way, the technology of astronomy having become immensely more powerful and sophisticated with the progress in computer technology. Nowadays, amateur astronomers can search for ET because the costs of computer electronics to process the observational data have become so modest. Something like a hundred and twenty amateurs in the SETI League have their own observatories devoted to SETI. And in the highly sophisticated SETI at Home program, which is run by the Space Sciences Laboratory at the University of California, Berkeley, there are a few *million* people who use their computers to search for intelligent signals that might be hidden in a vast mass of astronomical data regularly supplied by the giant Arecibo telescope in Puerto Rico. These millions of participants provide the world's most powerful computer-processing facility. Each participant regularly receives about thirty-five gigabytes of observational data to process. That's the first step. "When we get the data back . . . ," says Dan Werthimer, who directs the program, "we comb through the strong signals looking for ET."[2] No one has hit the jackpot yet, but SETI at Home makes it just possible.

There's more to the astronomical work than this. The detection of evidence doesn't depend on the presence of alien broadcasts. The detection of radiation that shouldn't be present, if the source were near enough to be recorded, might indicate the presence of alien technology (more on that later). However, all searchers for transmitted messages have a big problem: nature's immense timescales do not favor communications from one world to another—at least not by our speed-of-light level in radio and laser communications. The vast periods of time needed for the formation of suns and planets and the evolution of intelligent life make radio broadcasting an unrealistic means of gathering information about our cosmic neighbors. If advanced aliens or their intelligent robotic probes can travel to other planetary systems, they might not bother with broadcasting. The situation in our galaxy—some four billion years after the first civilizations evolved—may not be one where isolated worlds are transmitting detectable messages; this has been the predominant situation in SETI. Because the relevant science tells us that the first world civilizations could have started to emerge from about four billion years ago onward, those civilizations, if they existed, may have found us by now, and the best UFO reports might—just might—be an indication of this. However, good luck to the astronomers searching for transmitted signals, though they seem more likely to make astronomical discoveries than to discover the thoughts of ET. Yet even if they never receive a message from Planet X, their searches do promote large-scale public interest in the science of SETI, which is as relevant to our thinking about the human situation as anything could be.

A LIMITED PRINCIPLE

Since I hope to show why efforts to test the Grand Hypothesis are in need of modification, I have to refer to the principle of mediocrity, the validity of which had apparently been confirmed by our past understanding of our position in the universe. Nicolaus Copernicus started this line of thought in the fifteenth century when everyone was satisfied with the belief that we were at the center of the universe with everything orbiting the Earth, including the sun. Then the Italian philosopher Giordano Bruno realized there was nothing special about the sun and that all stars were suns, probably with worlds in orbit. This was too much for the Catholic Church, which by then had regrets about allowing Copernicus too much free expression, so it burned Bruno at the stake in 1600.

Galileo traveled along the same intellectual path toward the stake but wisely recanted his heresy and survived. Yet the belief that we were a special creation, separated by God from all other life on this planet, remained just about tenable until Charles Darwin. We then knew how we had arrived on this average type of habitable planet in orbit around a common type of star, and that we could have many counterparts on other worlds orbiting other stars. The principle of mediocrity had been fully vindicated—or so we thought. Thus the pioneering SETI scientists in the 1960s naturally extended the principle of mediocrity to our extraterrestrial neighbors. They would have their origins on Earthlike planets in planetary systems not unlike our own, and after about four billion years, life on those planets would evolve enterprising species that would develop civilizations—and radio. They had to invent radio. That was essential. They would then advance far beyond our level in science and technology but keep their old radio transmitters handy to contact new civilizations that had recently emerged from the Stone Age.

In appearance, those distant aliens might be shockingly strange, but that was all right as long as their brains worked like ours. This is not an unreasonable expectation if they evolved in similar conditions to ourselves and had to overcome similar problems to survive. Like us, they would have evolved a drive to communicate, since no species would rule its planet without good communications. And this drive to communicate would naturally extend into attempting to contact their cosmic neighbors. And for this, of course, they would use a simple technology everyone could understand and receive—basic radio technology. This was all very convenient for us because we already had the radio know-how to receive their signals and learn the secrets of the universe. But it was all taking the principle of mediocrity too far.

When we take a realistic look at this principle, we should realize that it cannot cover developments in the vastly different time frames in which other intelligent creatures will have evolved. They might once—and for a short time—have been radio enthusiasts, but they can't be expected to remain like that for long. While we are barely out of the cradle, technologically speaking, they will have been moving on. It's as certain as anything can be that we have had less time on the technological ladder than anyone else in our galaxy. This means that we will be different from everyone else at the moment. Not similar, as the principle of mediocrity would predict, but different. If the evolution of technological intelligence has been taking place for the past few billion years, as the astronomy of the oldest stars indicates as a possibility, we might even be embedded in a community of advanced civilizations without knowing it—like my cat and his feline friends that know nothing about anything outside their limited world. This may be sci-fi speculation, but something like it is not impossible.

Thus given that this brief review of the principle of mediocrity is somewhere near the mark, we could hardly expect a barrage of broadcasts beamed in our direction—or in any other direction. So what could we expect? If interstellar travel is possible—and no one can say it isn't—we might expect inexplicable phenomena that enter our world from time to time. And this is what the most credible UFO reports provide once all the hoaxes and wacky fantasy and genuine misidentifications are eliminated. Perhaps we should limit ourselves just to the reports from pilots and aircrew. Since the 1940s, there are about three thousand cataloged reports to consider. In addition, there are many reports from the military and the police, who are normally accepted as responsible recorders of events. But pilots and aircrews form the most reliable group because they are in the best position to see anything unusual in our skies. And although there are plenty of people on the ground whose brains regularly confirm the presence of UFOs and aliens, they won't be flying military or commercial aircraft.

Dr. Richard Haines, who was NASA's chief of the Space Human Factors Office, has been the main force in emphasizing the great importance of pilots and aircrews as witnesses. As he says, "They represent a very stable personality type with a high degree of training, motivation and selection."[3] Haines worked with pilots when carrying out research into their peripheral vision during takeoffs and landings. In the course of this work, he asked them if they had ever seen anything strange during their flights. "They have a unique advantage point simply by being in the air," says Haines, "if for no other reason than when the phenomenon is between your eyes and the ground you can calculate the slant range." This means that the position of the pilot, the object, and the ground can be used to estimate the distance and size of an object, something that cannot be done by anyone looking at an object against the sky. Sizes and distances are often given in UFO reports, although these can only be unreliable guesses unless a UFO is seen against a background of some sort.

"If a pilot comes forward with a strange tale," says Haines, "I give him a lot of careful concentration because he's putting his reputation on the line and maybe his job." After years of work, Haines assembled his air catalog, called AIRCAT, listing over three thousand cases reported by pilots and aircrews. Of course, only a small proportion might describe alien technology. Haines allows that the best in human vision can often be deceived. Stars and planets can seem nearby and can appear to move because of automatic movements of the eyeballs. "When you're flying in a sunny, clear blue atmosphere," says Haines, "the eye can focus inaccurately, so that you're not focusing at infinity anymore but maybe only one or two meters in front of the cockpit." Nevertheless, he maintains that these problems cannot account for all the cases in the cataloged data. "If a pilot describes a disk-shaped airform with no visible means of propulsion, pacing his right wing for 30 minutes—and I have plenty of cases like that—then that's not an optical illusion, it's not a bird or balloon or meteor, it's not any of those prosaic explanations."

And it's not only aircrews who have reported strange craft. Several astronauts have also mentioned some unexplained aerial phenomena. Prominent among these is Buzz Aldrin, who flew to the Moon with Neil Armstrong and Michael Collins in 1969. In a widely transmitted television interview in 2006, he described how a strange phenomenon flew alongside their Apollo spacecraft for part of the journey to the Moon. He and his fellow astronauts reported their observations to NASA, thinking what they saw might have been the remains of their last booster rocket, but they were told that this had been six thousand miles away. Aldrin explained that as their line of communication was open, they could hardly say that they had a UFO accompanying them on their way to the Moon. And later, of course, NASA quite understandably kept this information under wraps. But the Buzz Aldrin interview is on record, and his level of credibility is high, as is that of all the other astronauts who have reported strange aerial phenomena.

Many SETI scientists, however, still maintain that every UFO report would have a natural explanation if we had enough information, and the British government's UFO Study (2000 to 2004), released in 2006, cautiously supports this position. It shows how plasma balls could form in the atmosphere and sometimes even become attracted to the exhausts of aircraft because of the electrical charges they carry. The study says, "The charged mass is likely to be captured by the exhaust of aircraft engines and to follow the aircraft, maintaining both its shape and keeping a constant distance from the tail assembly." In other words, a sphere of plasma forms in the atmosphere, is attracted to the rear of an aircraft, and follows the craft, maintaining its position and integrity even though it is being dragged through the atmosphere at the speed of the aircraft—probably around four hundred miles an hour. The study doesn't explain how this is possible, but even if we accept that it does take place, this cannot explain all the cataloged cases, some of which involve reports of structured craft and phenomena that are not behind and following the aircraft. (Catalogs of UFO data from pilots and aircrews are freely available on the web, the main contributor being Dr. Richard Haines.)

It certainly seems that spherical plasmas do form in the atmosphere, and these may be reported as inexplicable objects by aircrews. But although this may explain some of the reports, it fails to explain them all. So it doesn't make scientific sense to dismiss apparently relevant data from such a credible group within society. What we need is more top-level research on atmospheric plasmas, a subject that only in recent years has been given proper attention.

However, even if all aircrew reports of UFOs could be explained in terms of natural phenomena, there are other possible sources of evidence. For example, scientists are just beginning to study photographs of the Moon and Mars for archaeological remains that could be millions of years old. Because of its proximity and lack of erosion, the Moon is the best prospect for finding evidence of any past visits. Anyone studying the Earth during the past billion years or so would almost certainly have visited the Moon. For much of Earth history, the Moon could have provided a good base camp—especially for intelligent robotic visitors. There are also some scientists preparing to search for alien probes in the solar system or in Earth orbits. Such research is on the fringe at present, but it doesn't go beyond the fringe. The scientists involved follow the rules of traditional science, while basing their research on the same rationale that has supported astronomical SETI for the past fifty years. There's only one difference: those who search for broadcasts from the stars deny the possibility of interstellar spaceflight. No matter how advanced our counterparts might become, they won't be visiting us—not even their highly intelligent robots, who, in theory, would be ideal space travelers. But those who dare to dabble in the diabolical idea that evidence of ETs may be within the solar system do not support such chauvinism. Unless you claim to know all the physics that might in future be applied to spaceflight, you cannot say that interstellar space travel is either impossible or too daunting for advanced civilizations. Major advances of science and technology look far more likely than unlikely, and interstellar spaceflight may not always be impossible. It may remain impossible for flesh-and-blood creatures like us, but not for the highly intelligent robots that we may developing in the future. And if this is a future possibility for us, others may have reached that stage a very long time ago.

A WORRYING SITUATION

For most people, the thought of having advanced biological aliens or their intelligent robots monitoring our activities is not exactly appealing. No matter how much people would like to know that life and intelligence is abundant in the universe, they would prefer to have their cosmic neighbors living light-years away rather than paying us regular visits. Many would go further than this, preferring humanity to be the sole intelligent inhabitants of the universe. Well, if not exactly the entire universe, then in our galaxy of a hundred billion stars. But this cozy view of existence is eroding. That life and intelligence may be universal is no longer dismissible speculation. Of course, proving a negative—that we are alone in the universe—would be impossible. We can never know that there have been no alien visits to the solar system or world civilizations signaling hopefully across interstellar space. We might prove the Grand Hypothesis, but we can never disprove it. So do we want to know that the evolution of life and intelligence are integral and inevitable in the great scheme of things, and that we are not the only beneficiaries of the ways in which the universal physics and chemistry work? We probably do want to know, because that is the way our brains have evolved. In the past, our ancestors needed to know things in order to survive. Nowadays it's usually a matter of curiosity. But one day that irresistible drive may take us, or our artifacts, to the stars and other planetary systems. It therefore follows that species that have created other civilizations should have evolved the same drive to know and may have long ago taken those trips to the stars that we at present can only dream about. So the drive to know and explore will be a part of highly intelligent beings everywhere, if they exist.

We could be alone, the only technologically intelligent animals to have evolved in the history of our galaxy. Life here could be the result of a string of chance events. We could be the jackpot winners of all jackpots, although what we know stacks heavily against life itself being a chance event. However, we do know that a series of chance events in the evolution of our ancestors did lead to the emergence of our technological intelligence in a recent period of Earth history. So it's still an open question whether ET is out there or not, although irrefutable evidence of ET technology in the solar system—or messages from the stars—would confirm that technological intelligence is universal. Of course, such evidence, especially if it exists in the solar system, would be a bit of a shock. Our view of the natural order of things would be changed forever. We would know that we must be the youngest player in something that we are yet to comprehend. Yet for this scenario to be true, our planet would probably have been known about for millions of years, so that visiting aliens are not about to take over the world. Neither is it likely that we would be able to engage in cozy communications. The *Star Trek* crews can converse freely with all the aliens they meet on distant worlds, but it could be nothing like that in reality. The rapid rates at which world civilizations might advance once they've reached our level of development makes the meeting of intellectual equals look impossible. For dramatic purposes, the *Star Trek* universe has many alien species, all at more or less the same intellectual level, voyaging in space with comparable technologies, but that's impossible. When we begin to touch reality in this subject—if we ever do—it's going to be very different.

THE ASTRONAUTICAL THEORY

A good example of applying traditional scientific methods to try to find out what this reality might be is Roy Dutton's astronautical theory, which he has developed during some forty years of part-time research. Dutton is an aerospace engineer, now retired, who was initially impressed by the flight characteristics of the strange craft that local people had reported. He was surprised that what the public was reporting could be explained as aerospace engineering. In the following years, as a complex theory evolved from the data, he found that his research was indicating the presence of some kind of monitoring system. This may seem an extraordinary result, but anyone who reads Dutton's research papers can see that the theory developed free of preconceptions on his part of what the UFO phenomena might be.[4] Some twenty years after starting his research, Dutton computerized his data, and the theory is now based on more than twelve hundred reports from many parts of the world; and like any other scientific theory, it will stand or fall by being tested. What it provides for testing—by radio or optical astronomers—are the coordinates of a system of advanced spacecraft in virtual orbits and virtual entry points from space. They are called "virtual orbits" because craft that enter these orbits would do so only transiently. Our satellites, by contrast, are always in their orbits until those orbits decay. But despite the detail the theory provides, it awaits testing by professional astronomers.

The following story illustrates the problem of getting such a radical theory tested. Roy Dutton and I had won a first prize of $5,000 for an essay in a competition run by the National Institute for Discovery Science (NIDS). Our subject was "Why ET Technology Might Be within the Solar System and How Could Scientists Test for its Presence." Dutton's contribution was a paper on his theory to test for the presence of probes. Mine was a review of the science that supports the hypothesis that there could be probes to detect. Subsequently, one of the scientists working for NIDS offered our essay to the British Interplanetary Society for publication in its quarterly journal. The editor was sufficiently interested to send it out to two referees. One referee pointed out that it was the policy of the society not to publish anything that mentioned the term *UFO*. The other referee said that an edited version might be published with the term *UFO* removed from the text. Our essay was never published in the journal, although NIDS posted it on its website. Dutton felt so thwarted by this experience that he replaced *UFO* in all his research papers with the term *SAC* (strange aerial craft).

WHAT'S OUT THERE?

Dutton's theory illustrates the point that since we don't know what our counterparts might be like or what they might be doing, the broader the approach to SETI, the better. The evolution of technological species may be very rare, judging from our own precarious evolution. Millions of years may separate the emergence of technological civilizations in our galaxy, and thousands of light-years may separate their planets of origin. Thus the exotic transport enjoyed by *Star Trek* crews, with all the comforts of a good hotel and humanoids providing hospitality on every other planet, is not something we should count on. If thousands of light-years separate the centers of civilization in our galaxy, then radio or lasers or, far more likely, some superior means of communication may be the only way to tell others of one's existence. If advanced physics could offer a basis for rapid or instantaneous communications, rather than waiting hundreds or thousands of years for the post to arrive, then no one would need to visit anyone, providing everyone had the necessary communications technology and a reliable address book. Alternatively, instantaneous communication across the light-years may be impossible. The speed of light may be a barrier forever, which would certainly discourage interstellar broadcasting, especially if interstellar space travel enabled civilizations to spread into neighboring planetary systems. The planet-hunting astronomers have found plenty of planetary systems where super civilizations might set up colonies. Hundreds of planetary systems exist in just our tiny region of the galaxy, where their presence is near enough to be detected with current technology.

In the coming decades, astronomers will scan some of these systems to record the spectral lines that would show the presence of life. So if only one planetary system in a hundred has a life-supporting planet, that is the one we would send probes to (if we had the probes to send), not the 99 percent where no life exists. And we might expect highly intelligent aliens to do the same. So it's not impossible that "they" may have worked their way across the galaxy, from one planetary system to the next and so on for many millions of years. They may not have, of course, but in theory the oldest sunlike stars could have shone upon space-age civilizations four billion years ago, so there's been plenty of time for intelligent beings (biological and artificial) to have found the solar system and our planet. There's been enough time for every life-supporting planetary system in the galaxy to have been visited, providing interstellar travel is possible—at least by intelligent robots. Flesh and blood may never make it to the stars, but intelligent robots would make ideal space travelers.

Keep in mind that for most of a planet's history there would be no intelligent life and civilizations to study. A look at Earth history suggests that a high proportion of planets with thriving biospheres may never evolve creatures capable of creating anything but the next generation. In our case, visitors would have found only microbes until a billion years ago, and not much more for the next four hundred million years. Alien visitors might well have found the biochemistry and genetics of Earthly microbes fascinating, but the past few hundred million years have produced jellyfish, real fish, giant amphibians, and reptiles, plus a fine lot of furry mammals during the past fifty million years.

Yet but for a fortuitous set of environmental events in the past few million years that boosted prehuman evolution, alien visitors would never have seen civilization on Earth. "Flourishing biospheres" are one thing; "planetary civilizations" are another. So let's indulge in a little sci-fi speculation. Could it explain why there are visitors in our epoch when the Earth's spectral lines have been signaling the presence of life here for more than 350 million years? It might, though that's really a side issue. We first have to test for their presence in the solar system, either by looking for past relics on the Moon and Mars, as some scientists have done, or by testing aspects of the UFO phenomena that may or may not have an extraterrestrial connection.

EARTH WITHOUT HUMANS

In 2004, a BBC television series depicted the next two hundred million years of evolution on Earth. Prominent evolutionary biologists contributed, speculating on what might happen if *Homo sapiens* became extinct and left the planet free for the unrestricted evolution of other animals. They showed how these might evolve as environments changed—and environments are always changing. But these biologists never showed how a new technological species could evolve, even though they had two hundred million years of evolution to play with—and the fact that we managed to evolve from squirrel-sized primates in fifty million years. Of course, if the BBC's participants had evolved a technological species after another fifty million years, all subsequent life-forms would have had their evolution restricted from then on. So to keep the series going with plenty of new animals, the producers were wise to avoid the evolution of a new technological species. And by doing so, they avoided the problem of finding a life-form in the animal kingdom that, with the right evolutionary pressures, could have provided a replacement for *Homo sapiens*.

This would not have been easy, and we have to concede that it might never happen if *Homo sapiens* became extinct. The Earth might be left to an amazing lot of dumb animals until its dying days.

Consequently, if you're a super extraterrestrial concerned with galactic culture—and you have a good supply of flying saucers—it's those rare planetary systems with civilizations that you'd keep an eye on. This sounds like science fiction, and it probably is, but it's not inconsistent with the worldwide reports of strange aerial craft (SACs, in Dutton's terminology). Our rapidly changing world would be far more interesting today to alien eyes than at any other time in Earth history. But that doesn't mean they're here.

PROPULSION CHAUVANISM

Could the aliens get here? That's the question. Obviously, if you are searching for signs of intelligence among the stars, you would rather not have flying saucers hovering overhead. So the barrier of interstellar space against such discomforting visitors is firmly fixed in the minds of SETI astronomers: "We can't see a way of visiting them so they can't visit us. No doubt about it." However, the universe has to be rather short of applicable physics for us to put such a definite limit on spaceflight. It shows rather more than a trace of chauvinism. Way back in 1963, a collection of SETI papers was published as *Interstellar Communication* by Benjamin Press in New York. One of the papers, by Edward Purcell, then a professor of physics at Harvard University, reviewed the problems of interstellar travel to show that it was not going to be possible—ever. The paper was written in 1963, but the views then expressed still linger.

Even in the journal *Science* in 1992 we find a prominent member of the SETI community, Ronald Bracewell, saying, "The idea of interstellar travel by intelligent beings was debunked by Edward Purcell."[5] He also said, "Frank Drake does not like the idea of a brain-sized interstellar probe . . . partly because of the large retro-rocket that would be needed to bring the probe into orbit in the targeted planetary system." Clearly, that's only reasonable if "large retro-rockets" were needed for "brain-sized probes." The point is that by the time people can build "brain-sized probes," they may have an equally advanced propulsion system, leaving "large retro-rockets" to technological history.

An old quote from the late J. Allen Hynek, a professor of astronomy and pioneer UFO investigator, puts such comments into perspective: "I have begun to feel that there is a tendency in 20th Century science to forget that there will be a 21st Century science, and indeed a 30th Century science, from which vantage points our knowledge of the universe may appear quite different than it does to us. We suffer, perhaps, from temporal provincialism, a form of arrogance that has always irritated posterity."[6]

What Hynek said may make us more realistic about the situation in SETI. Gerry Zeitlin of Open SETI has provided a good example of Hynek's viewpoint. He quotes Dr. Ragbir Bhathal of the University of Western Sydney, who says, "The National Ignition Facility in the US has produced laser powers in the teraWatt range . . . albeit for short periods. These developments give tremendous credence to the search for ETI [extraterrestrial intelligence] signals in the form of nanosecond laser pulses."[7]

Unfortunately, these amazing developments in our laser technology can't give such credence in the vast time frames of planetary development and biological evolution that are the backdrop to SETI. Laser technology on Earth a thousand years from now—if not replaced by something else—will be quite different from anything the brilliant minds at the National Ignition Facility are perfecting. The lifespan of any technology can be quite brief before something better replaces it, so SETI scientists will always have this problem of the "technology gap." One way to avoid it is to base research on reported data, in the case of the UFO phenomena, or to look for spectral lines that shouldn't be present from bodies in the solar system. Such lines won't transmit "messages," but they might indicate the presence of alien technologies at work. The nature of such evidence, originating from the nature of chemistry and physics, would at least not change over the millennia.

Such research has already been pursued by one astronomy professor, the late Michael Papagiannis, who looked for spectral evidence of tritium from alien probes; tritium being a short-lived by-product of nuclear fusion. This sort of searching might eventually be carried out on some of the newly discovered planetary systems whose suns are comparable in age to the sun and from which the spectral lines that indicate the presence of life have already been detected. It's a job for the future, but it might be done one day.

LOOK OUT FOR ROBOTS

Having said all this about the need for a nonspeculative approach in the SETI-UFO arena, I would like to offer a little speculation about the probable future of artificial intelligence in space. I find it hard to foresee biological beings voyaging through space like the *Star Trek* crews and encountering the humanoid aliens they are always bumping into as they cruise across the galaxy. Apart from the highly exotic physics that would be needed to make regular trips to the stars, it would be exceedingly difficult to maintain the complex support systems for biological beings light-years away from home—just another big problem on top of the big problem of perfecting an interstellar propulsion system. And it could be avoided. So, going back to *Star Trek*, my guess is that a less entertaining but more realistic scenario would have only intelligent robots aboard all those spaceships, those belonging to us and to the aliens. This is not wild speculation. Scientists developing artificial intelligences already talk of the "singularity," a point in the near future when artificial intelligence will equal or better human intelligence, which could pose big problems for *Homo sapiens* before this century is out. Yet it's not just a matter of intelligence. We have qualities and abilities beyond pure intelligence. We are highly efficient biological entities, and it has taken almost four billion years of natural selection to create us, whereas the history of robotics is less than a hundred years old.

However, by the time interstellar transport is perfected, everyone may realize that robots are the obvious candidates to explore the galaxy, and that biological creatures like us who need complex life support should not think of ever leaving their planetary systems. The robots wouldn't need air, central heating, water, food, or toilet facilities. They wouldn't breathe out carbon dioxide or produce any waste that needed to be recycled. And they wouldn't have problems with their kidneys, lungs, liver, or heart because they would have this biological baggage that evolution has necessarily given us to keep us going. In short, artificial life would be completely free from the complications of biology. Robots could switch off for hundreds of years at a time, traveling unconscious from planetary system to planetary system, just waking up in time for the interesting parts of a cosmic journey. As far as I know, sci-fi films haven't shown us that future. The visiting aliens are always flesh and blood. Their blood may be green, but they're still flesh and blood and as vulnerable to the hazards of life as we are. The Martians in *The War of the Worlds*, the great sci-fi novel by H. G. Wells, had the technology to come to Earth, and to conquer us, but lacked the biological knowledge to stay healthy after they arrived. They didn't know about bacteria, a fatal flaw in their knowledge and very convenient for Wells, who might not otherwise have gotten rid of the Martians.

Of course, real flesh-and-blood aliens could be better prepared, though why not leave it all to the robots who couldn't be infected by any biological bug on any planet? Judging from the way our technology of artificial intelligence is developing, we could one day have our own robotic astronauts who, like the android Data on *Star Trek*, might not be easily distinguished from real biology. They might be near immortals that could cross a thousand light-years and arrive in pristine condition, having slept electronic sleep most of the way. And if we can perceive this as possible, then older and more advanced civilizations may have created the ultimate in robotic life long ago. I haven't seen the film yet where robotic life and wisdom encounter an irrational human society on a beautiful blue planet called Earth, but any interested film director should contact me before the aliens arrive.

ROBOTS INCREASE THE COLONIZATION FACTOR

Stephen Webb, a good writer on the subject of SETI, has written in his *If the Universe Is Teeming with Aliens . . . Where Is Everybody?* that "if the number of potential life-bearing planets is much smaller than most estimates suppose, then the number of potential extraterrestrial civilizations out there must also be smaller." I think the reader may see that this does not follow. This is an aspect of SETI that is not often allowed for. The point is that the science and technology of artificial intelligence has advanced enormously in such a short time. Specialists working in this subject already envisage robots with a human level of intelligence as a certain future development, and this probability drastically affects any estimates of the number of technological civilizations in the galaxy. While intelligent robots would be ideal space travelers, as we've seen, the requirements of biological beings would not be easily met anywhere beyond their home planet.

However, robots could colonize planets without atmospheres and water and having a temperature range too high or low for biological space travelers. So, in theory, there could be a lot of artificial intelligences out there inhabiting planets unsuitable for organic life, but whether or not they would want to contact us is another question. Perhaps they would wait until we have created a population of rational robots. Or perhaps they would prefer worlds completely free from the interference of troublesome biological creatures. We don't know what the situation is, but we can see that the door to the exploration and colonization of the galaxy becomes wide open once a high level of artificial intelligence is created. No need to maintain biological life throughout long interstellar journeys when robots could take hundreds or thousands of years to reach the next planetary system at speeds well below that of light.

Of course this doesn't correspond with the old idea of space scientists and science fiction writers in which humanity boldly goes forth to explore other worlds and other life. But it's a realistic prospect, whereas the old ambition is not. To be realistic, the *Star Trek* crews should all be robotic androids like Data, programmed as he is with the best of human qualities. As well, all those races of space traveler that they keep bumping into in the vastness of space should also be robotic. The only alternative to this rather pessimistic scenario would be the creation of some currently inconceivable means of crossing interstellar space. The knowledge to do this may be hidden in the heart of nature, but we can't see it yet.

BETTER ROBOTS BY
NATURAL SELECTION?

It has been suggested that radiation in space or on an unprotected world could change the programming of robots, as radiation has changed biological organisms when their DNA has been damaged or altered. It's from damaged DNA and mistakes in copying that life has evolved more complexity and variety to suit conditions on Earth, though only a tiny proportion of such changes to the genetic makeup of organisms are likely to be beneficial.

Not so for robots. Advanced robots are not going to depend in this way on Darwinian evolution. They could repair any damage to themselves and immediately improve themselves and their next generation. No need to wait a few million years for natural selection to do a job they could do in a day. And once they reach the "takeoff point" with an intelligence equal to ours, or greater, they could soon start to develop themselves into the masters of everything that attracted their interest, including the colonization of the galaxy.

Yet even the realistic prospect of robotic space travelers does not persuade SETI astronomers that evidence of alien technology might be within the solar system. They continue to scan the galaxy and stand by their assumption that other world civilizations won't take on interstellar space—not even the robots. The aliens will all stay at home for millions of years and, what is more convenient for us, use radio technology at our level to contact emerging civilizations. Why? To bestow their knowledge and wisdom upon us, of course. This belief in big-brained and big-hearted aliens who would make things easy for newly hatched technologists was once accepted by people enthusiastic about SETI, including myself. But I awoke from that dream.

In the early days of SETI, we accepted more than was reasonable because some of the brightest people around established SETI and convinced us—and themselves—with their clever ideas. Yet in the heads of many SETI scientists the dream continues. For fifty years, in dozens of observational programs, only a few SETI astronomers have deviated much from the original line of thought. Most still search the same narrow band of frequencies, though there have been a few enterprising departures in the official ranks. One of these was Michael Papagiannis, who, in addition to searching for evidence of probes by looking

for the main frequency of tritium, studied two thousand spectra from asteroids recorded by the Infrared Astronomical Satellite. He wanted to see if any
showed spectral characteristics not consistent with asteroids.

Two other astronomers, Robert Freitas and Francisco Valdes, were earlier
involved in similar research, using a large optical telescope in two research programs to look for alien probes parked in the so-called Lagrange points of the
Earth-Moon gravitational system. These are stable gravitational regions where
visiting aliens could conveniently park their probes, knowing that gravity
wouldn't move them much in the next few million years. These searches
for probes deviated significantly from mainline SETI, yet this research was
accepted by the science community because good scientists were involved. Of
course, Freitas and Valdes were searching for ancient probes—or so the science
community assumed—and that made a difference. It's just about acceptable
to search for ancient alien probes, but they have to be really ancient and inac-
tive. If they become active probes, or "UFOs," or, even worse, "flying saucers,"
alarm bells ring, calling all scientists to reject such unhealthy ideas. Yet those
three astronomers set a precedent that shows a conspicuous inconsistency in
mainstream SETI. They were searching for probes that had crossed interstellar
space to the solar system, as any alien craft would have to do.

To my knowledge, only one observatory has searched for alien probes since
then, although the theoretical basis for their existence is probably sounder
than for alien broadcasts, a conclusion that is just beginning to reach the SETI
community. At least one prominent member allows it as a possibility. Dr. Jill
Tarter, a pioneer in SETI astronomy and director of SETI research at the SETI
Institute in California, was quoted in the *New Scientist* as saying, "They could
be here. I don't mean they are abducting people off the street, or landing on
the White House lawn. But we have so poorly explored our own Solar System,
there could be probes here that we don't know about."

If Jill Tarter thinks like this, then it is unlikely that aliens travel here
simply to park themselves permanently in the outer regions of the solar system,
never bothering to explore Mars, the Moon and the Earth. It could hardly be
like that. If they came in the distant past, then evidence may still be on the
erosion-free Moon and on Mars. So Dr. Tarter and other SETI scientists have
more or less implied that we could explore the solar system for probes, yet at

the same time they want to ignore any anomalous phenomena (strange aerial objects) that could be associated with advanced probes in Earth or solar orbits. And those scientists who are looking for signs of ancient artificial structures on the Moon and Mars are not exactly acclaimed as enterprising investigators by the science establishment. Imagine the possible scenario: while teams of scientists scan the galaxy, looking for the signature of alien broadcasters, another group of researchers find photographic evidence of decayed buildings on the Moon. The mainstream SETI scientists, and many other people, think this is impossible because no one has ever crossed interstellar space to reach the solar system. So looking for unnatural features on the Moon, using NASA's best available photography, is very much a fringe area for research.

We can see from Jill Tarter's comment how the colorful aspects of ufology will find no place in scientific research. Checking on the existence of probes is best left to physical scientists and should be carried out for the reasons we've considered. Checking on the reality or otherwise of abductions, which she mentions, has nothing to do with this. The abduction phenomenon is work for forensic scientists and psychologists. And since no forensic report has yet confirmed the presence of any extraterrestrial evidence on the clothing of any abductee, and since no abductee has ever had anything of scientific significance to tell us or show us, the phenomenon looks like work for psychologists. The people who claim to be abductees may be sincere and normal, but we can't believe them unless they can produce evidence. And it needs to be substantial evidence because what they claim would verify the major hypothesis in science. But we'll examine the abduction phenomenon later.

ONLY ONE CERTAINTY

Only one thing seems certain about this subject: if ET technology is in the solar system, it's been here an eternity. Our present awareness of UFOs, whether they're structured craft or plasmas or inexplicable blobs of light, results from the growth of air traffic, radar, and worldwide communications. Geologists and biologists tell us that a highly evolved biosphere existed on Earth from about 350 million years ago when dry land was first being colonized. They haven't, however, made much of the fact that it would have been relatively easy, from that time on, for advanced worlds in our region of space to have detected evidence of life on Earth. Projects are being developed today by our astronomers to detect life in the nearest planetary systems, so we can reasonably assume that other advanced civilizations would have done the same long ago.

The fact is that highly evolved biospheres will display the spectral lines of life, including the prominent spectral line of ozone, where photosynthesizing life has produced enough oxygen to form an ozone layer. It's something of a gift from nature that the universal physics and chemistry makes it possible for intelligent life to find life in other planetary systems, providing that life is near enough. As a consequence of this, the Earth has been an interesting target for more than 350 million years, and our thinking about alien probes and the possible detection of the signatures of life should take this into account.

Societies such as the Society for Planetary SETI Research, which looks for photographic evidence on the Moon and Mars, and the Organization for SETV Research, which is interested in finding evidence of alien probes, have as their rationale the possibility that the solar system has been visited. The SETV group of scientists and engineers, which had its origin in NASA's famous Jet Propulsion Laboratory, make a good case for alien probes being in Earth or solar orbits. But they carefully avoid contamination from the imaginative side of ufology.

In its manifesto, the SETV allows that only .04 of all UFO reports may indicate an ET presence. This is a strangely precise figure with no mention of how it was calculated. However, with so many UFO reports available, it does represent a large number. To be conservative (an excuse for making the arithmetic easy), let's take only the thousand most credible UFO reports. Now, .04 of 1,000 is 40, so, according to the scientists in the SETV group, at least 40 UFO reports indicate the presence of alien technology. That's enough to warrant serious scientific attention. And such serious scientific attention is relatively inexpensive, whether it takes the form of searching for probes in orbit, looking for ancient alien sites on the Moon and Mars, or studying the most credible aspects of the UFO phenomena. This research can make use of freely available data, including some from the space programs. Nevertheless, the level of proof would have to satisfy the top brass in the scientific community, say, the Royal Society in London and the American Academy of Sciences in the United States, and comparable scientific societies in other countries.

THAT SOMETHING SPECIAL

We can see by now how much of SETI is based on informed speculation with hardly any data in sight, so it's difficult to proceed in a normal scientific manner. In astronomical SETI, you can show that no aliens were broadcasting on a given frequency during the brief period that a given region of space was being scanned. But that's about it. Signals might be arriving continuously for all we know, but they come from technology we haven't yet invented. Astronomers can't know if they missed a good broadcast that ended a million years ago if their equipment is too primitive to catch the latest galactic news. Yet even if civilizations out there were using similar systems for their communications, there's still the problem of catching anyone at home during the past few billion years. Remember that bacteria ruled the Earth for the first two-thirds of life's history, followed by minor creatures for much of the last third. Only during the past 350 million years have large land animals evolved, and none were noted for their radio technology. SETI astronomers, by comparison, have been around for only fifty years, and no distant civilization would get a reply from planets devoid of astronomers. All bright aliens will know this. Only instantaneous communications across the light-years could make broadcasting the preferred option, but in that case we haven't got the right technology to chat with our cosmic neighbors.

HOW MANY?

It's an impossible question to answer, but SETI scientists still try to estimate how many civilizations might be transmitting intelligent signals. If the number is really enormous, then there might be one backward world like ours that was, or is, broadcasting in the old-fashioned way. (Remember that detectable signals could come from anywhere in our galaxy, which is a hundred thousand light-years across. So signals received now could have been transmitted many thousands of years ago. How long ago would depend on the position of the sender.)

SETI scientists have published a library of guesses on the number of transmitting civilizations, using several factors—mostly unknown factors— to estimate the number of such civilizations. However, our own history shows that once technological intelligence has evolved, the rate of development can explode with no end in sight. Also, we can see that the evolutionary process to reach our position can be extremely precarious, and that something more than the emergence of toolmaking is needed.

Our direct ancestor *Homo erectus* had a brain two-thirds as large as ours, yet anthropologists tell us he was slow to improve on his toolmaking in a million years. Contrast this with what *Homo sapiens* have done in a few thousand years. Something clicked in the brains of our ancestors, which enabled them to see ways of continually improving their technologies. That has only happened in one species in the entire history of life on Earth. The Neanderthals were our closest relatives with brains on average slightly larger than ours, but they, too, were not exactly toolmaking entrepreneurs. This indicates that creatures who create advanced civilizations on other worlds will have had an evolution that forced the development of intelligence and gave them that "something special" in their brains that we have and that drives us to improve on what we already possess. That "something special" is a unique asset on Earth, and only a tiny proportion of Earthlike planets may have a life-form that possesses it. So the abundance or otherwise of technological life-forms is a fundamental uncertainty.

Nevertheless, the optimistic line of speculation is saved if we accept that a few rare and long-lived civilizations have evolved during the past four billion years or so. Such great civilizations may have developed advanced artificial intelligence and interstellar space travel and spread from planetary system to planetary system, eventually reaching the solar system, perhaps drawn to it by the Earth's spectroscopic life markers. That's possible, but we can't begin to guess what has happened until we have some evidence, which may or may not be present.

WHY NOW?

Besides the fact that every odd event tends to be noticed by the media these days, the sci-fi scenario would be that alien visitors are here today because our planet is becoming an increasingly interesting place. Take air travel, for example. Suddenly, after ten thousand years of civilization, we start to fly. Vast numbers of vehicles become airborne. It's enough to make any alien take a close look—which would explain all those encounters with civil and military aircraft. For most of Earth history, one million years looked like the previous million years: same landscapes, same life-forms, no sudden developments to surprise anyone. Not exactly a tourist attraction. But these days not a decade goes by without major changes on the planet's surface. Add to this the unlikely event of *Homo sapiens* evolving from those vulnerable upright apes, dodging extinction through ice ages and other catastrophes, and we can guess that our civilization would be a rather unlikely spectacle.

If aliens exist with the ultimate in space technology, our brief period of history would be the best time to visit and maybe to establish bases in the solar system—just to see if we're going to break through the insanity barrier and make ourselves a permanent fixture in the galaxy. This may be sci-fi speculation, but it's one way to account for the many reports, photographs, and videotapes of strange aerial craft in our skies. Yes, much of the so-called evidence is bogus, but is it all bogus? That is the question. And it's a really serious question. Being the object of study by visiting aliens, or their probes, is decidedly different from receiving wisdom and knowledge from benevolent broadcasters on the other side of the galaxy. I'm speaking here of SETI's holy grail, the *Encyclopedia Galactica*.

Over the years, I've concluded that SETI astronomers will never receive the wisdom of the galaxy beamed out across space to aid new world civilizations. Once upon a time at international conferences, they talked about this with naive enthusiasm. Anyone can check this far-fetched expectation by reading the conference reports from the 1960s onward. It was a fine dream, and I shared it for a while, but some of us now accept that aliens will not be in the education business. A responsible civilization would leave us to make our own discoveries—and mistakes. They'd know that giving new knowledge and power to recently evolved technologists could lead to extinction.

Like most people in a brand-new endeavor, SETI scientists were too optimistic in their early days. Yet they have pioneered the search for universal life and intelligence, an achievement that has led to a valuable synthesis of science. This forms the SETI rationale from which we can draw support for the hypothesis that life and intelligence are universal. And the fact that the science establishment has accepted astronomical SETI as a respectable activity is a tribute to this and to the technical sophistication of its ongoing research. The supporting synthesis of science is also a major contribution to our understanding of the human situation on this planet, even though SETI's searches have been based too much on hopeful speculation.

One example is the narrow band of microwave frequencies covering what has been called the "water hole." The water hole band was so named because it has the frequency of neutral hydrogen (H) at one end and the frequency of the hydroxyl radical (HO) at the other. (A radical is part of a molecule.) On paper (but not chemically), you can put these two together and make water (H_2O). The idea behind this is that animals everywhere in the galaxy would come to drink at water holes, and that intelligent aliens would recognize this as a meeting place, whatever they might call "hydrogen" and the "hydroxyl radical."

There's more to the water hole idea than this. Its frequency range exists in the quietest section of all radiation reaching the Earth. So the "water hole" was a clever idea, even though the aliens do not appear to be thirsty at the moment. However, most SETI astronomers are not discouraged by this and continue to monitor frequencies within the narrow water hole band, including the main hydrogen frequency that is at one end of it. Of course, this is second-guessing the aliens—and it's good fun second-guessing the aliens. But when we have no idea what the aliens might be doing (if they are doing anything), it could be a mistake to back only one narrow line of speculation. Most SETI research has tested the hypothesis that the aliens are broadcasting on the water hole band of frequencies, while reports of inexplicable phenomena that do not fit our present paradigm have been ignored.

This approach fails to acknowledge that we do not know what may be possible and that we do not know the extent of applicable knowledge. Nature may offer a virtually limitless range of knowledge to support new technologies. The ceiling could be very high or very low. That is, we may have already bumped our heads on that ceiling and discovered most of the science that could be applied. If we are already near that low ceiling, then a lot of hopeful thinking in SETI becomes more credible. All civilizations may be stuck in their own planetary systems with radio technology the ultimate means of communication. Alternatively, and far more likely, scientific knowledge is virtually limitless. Given this is so, the scientific basis for alien technologies could be beyond the human brain's capacity to comprehend. The next stage of intelligent life, *Homo super sapiens* or even *Homo roboticus*, may be needed to do so. If the end of exploitable scientific knowledge is nowhere in sight, then other planetary civilizations could have spaceships that cross interstellar space as easily as we cross continents. They might not, of course, but strange and incomprehensible phenomena in our environment might indicate that they can.

This is where the "local" approach to SETI could have significance. It is looking for the signatures of life in a way that could move us toward a new paradigm. Finding a lonely intelligent signal from a distant part of the galaxy would only confirm what the majority of people already believe. The sci-fi sagas have prepared the world for this. But if we confirm that some UFOs (SACs) are part of a system that is monitoring our civilization, it would show that our position is distinctly different from what we've always assumed. *Homo sapiens* would not after all be in charge of the universe, yet our lives on Earth should be unaffected. All the relevant data points to that conclusion: there is no evidence of any contact in geological or historic times. There would be no danger of a takeover. But *Homo sapiens* worldwide might begin to feel a closer attachment to other *Homo sapiens*.

The factors in our favor could be that flourishing biospheres like ours are probably very rare and planetary civilizations even rarer. That's what it looks like from the way we've evolved. And because the time frames of evolution are so vast, the present state of our developing civilization should be unique in the galaxy. There will be no one at our stage of development at the moment, unless the ceiling to science is very low and we are already at the peak of technological development. So why would the ETIs want to end their study of such a rare and interesting subject by bringing chaos to Earth?

I was once on a platform answering questions alongside a civil servant who had been the British government's official spokesman on UFOs. One questioner asked, "What should be done in response to the continuing UFO presence?" The civil servant thought we should prepare ourselves for any unwelcome contact. I hope government scientists realize that this would be both pointless and unnecessary. The fate of our planet is not at stake from aliens. It's *Homo sapiens* I'm not so sure about.

CHAPTER 4

THE LIFE OF ALIENS

The young Captain Kirk's habit of kissing glamorous female aliens on *Star Trek* could have gotten him into a lot of trouble. Not only might he have acquired a collection of troublesome alien microbes, but he could have transferred microbes from Earth's biosphere to the aliens. A bacteriologist specializing in dental research once told me that there are about nine hundred different species of bacteria living in their billions around our teeth and tongue. But they're no obstacle to romance on Earth because they've been with us since our Stone Age days. Almost all are beneficial at best and harmless at worst, but they might not take kindly to romantic aliens.

In a real-life *Star Trek* situation, the problem wouldn't arise. No one would want to kiss the aliens. Nor would the aliens fancy kissing Captain Kirk, who might seem to them to be horribly weird. Thus the incompatibility of biospheres, from microbes to the most advanced biological creatures, looks like a permanent social barrier as well as a biological one. The hospitality our space travelers have enjoyed on alien planets in the sci-fi sagas is confined to the cinema and television. All biological aliens are going to be complex living ecosystems like us. They couldn't live otherwise, and they couldn't have evolved otherwise. We have evolved so that bacteria live within and upon us and keep us alive and healthy. We would die without them. We carry around more bacteria as part of our digestive system than we have cells in our entire body. The various species of bacteria live in their billions as a major part of our own personal ecosystem. It's the same for all animals. Thus a science journalist, after seeing Steven Spielberg's film *Jurassic Park*, wondered how the dinosaurs, which had only been reconstructed from their DNA, managed to live in today's world without their original bacteria. The bacteria that flourished in the guts of dinosaurs up to about sixty-five million years ago would no longer be around to help digest a dinosaur's dinner and keep it alive.

This interdependence of life upon life is found everywhere on Earth, so we'd expect the same setup to exist in other planetary biospheres. If so, only nonbiological space travelers, robots like *Star Trek*'s Data, would not be walking ecosystems. Highly intelligent robots could live and work freely in other planetary biospheres—a frustrating future for *Homo sapiens* wanting to visit those "brave new worlds." However, we can't ignore the fact that *Homo sapiens* are laden with biological baggage that makes long space voyages and life on other worlds impractical. For us, the robotics revolution is only just beginning, but for economic, social, and political reasons, no one can stop it. Advanced robots without our biological needs and limitations will be with us one day, and who better to crew future starships from Earth? And this would have applied in alien worlds where intelligent creatures evolved long ago. So if a flying saucer lands on the White House lawn, or anywhere else, we shouldn't expect to see the alien equivalent of Captain Kirk. More likely it would be the equivalent of the android Data but looking like the original biological masters of its home planet.

Such visitors might seem unlikely, but we could find some real aliens on Mars and in the ocean beneath the icy surface of Jupiter's moon Europa. The Martians would be microbes, but something larger might live in Europa's ocean. If these microbes exist, the big question is, are their genetic arrangements and codes drastically different from the genetic system and code that operates within all life on Earth? The point is that our code, since it has evolved here during the past four billion years, must be unique to our biosphere. We're so sure about this that if a similar genetic code were discovered in microbes on Mars, it would mean that those microbes and life on Earth were related, and that at some time, perhaps at the dawn of life when large meteorites were hitting the Earth and throwing rocks into space, our bacteria went to Mars encased in rock. We know that bacteria live within rocks and are real tough cookies, so some may have survived millions of years in space, protected from radiation, to eventually land on Mars. This explanation would make sense if Martian microbes had a similar genetic system to life on Earth. Alternatively—and perhaps more likely—it could have happened the other way around. Meteorites hitting Mars, where gravity is a third of ours, might have sent Martian bacteria on a trip to Earth.

Of course, if Martian life was found to be distinctly different at a molecular level from life here, it would demonstrate the options life has at a basic molecular level. In recent decades, scientists have discovered the marvel and complexity of our genetic system and code, which started to evolve about four billion years ago. And some have begun to think about how different genetic systems could be on other worlds. Has evolution provided us with something like the best possible system, or could it be better?

Anyway, because alien microbes might function in ways that could harm us, we have to prepare for any possible contamination. The first alien microbes to arrive—if any do—will come from the robotic spacecraft returning samples from Mars—and no one knows what damage they might do. NASA is therefore going to spend millions of dollars to quarantine the samples when they eventually arrive. Special facilities are planned to protect us, though NASA scientists believe there may be no danger, even if the Martian samples contain live microbes. But precautions have to be taken.

Of course, if UFO contactees and abductees have actually been with the aliens, as they claim, or if anyone has dead aliens hidden away in a freezer, then NASA is wasting its time and money. Alien life will have already entered the Earth's biosphere with a good supply of microbes unless everything associated with the UFO phenomena is entirely robotic and sterile. But since no contactee or abductee has ever reported as much as an attack of alien flu, the world can relax about being contaminated by alien microbes—or can it? Maybe the NASA scientists, who normally function on the basis of evidence, don't believe the claims from ufology.

Let's look at the science. In the best-case scenario, alien bacteria might not flourish on Earth because the food our cells supply, if we were infected, would not contain the right set of amino acid molecules they needed for their microbial metabolism and growth. They couldn't make their own proteins. Genuine alien bacteria would have evolved different systems of genetics and would use different genetic codes to make their proteins. Nevertheless, we can't be sure that they wouldn't find something in a human body to nourish them, something from which to synthesize everything they needed.

Alien viruses should not pose the same kind of problem, although if microbes flourish beneath the surface of Mars they could be hosts to viruses, as our bacteria are. Evolution does not waste an exploitable opportunity, and bacteria are large enough to offer a good home for viruses. So if we bring back live Martian microbes, we might bring back their viruses as well. Viruses are the perfect parasite, but they need the right host to flourish—they are very specific parasites. So we might not have a problem with Martian viruses. Our viruses have to be able to enter the cells of their hosts and use the genetic system there to make copies of themselves. They don't reproduce themselves like all other organisms in the biosphere; they get the host to do that for them. They are therefore entirely dependent on being able to enter the host's cells to use its genetic machinery to make more viruses. In order to gain entry and do this, the structure of the proteins on the virus must be able to key into the structure of the proteins on the host cells. This is the viral way of life, and it is very finely tuned. For example, I don't catch a cold from my cat when he sneezes, and as far as I know, he's never caught a cold from me. So viruses are mainly species specific even within one biosphere—our biosphere—although some do cross the species barrier. We do catch some nasty viral diseases from other mammals and birds, but our great biological differences from aliens could provide an uncrossable barrier for their viruses. Nevertheless, the vast abundance of microbes that would probably provide the basic organic support for all planetary biospheres—as our microbes support our biosphere—might be a problem if *Homo sapiens* ever visit other inhabited worlds.

By the time interstellar travel is possible, perhaps within a few centuries, there might be plenty of nonbiological astronauts ready for takeoff. That's a scenario waiting for a film producer. A cast of adventurous robots might lack audience appeal, but their credentials for space travel would be impeccable: the ability to travel without air, food, water, or toilet facilities—and to resist all microbial infections. Also, no need to study operating manuals or learn anything. If you're a robot, just download the required data to your robotic brain. Robots could work without rest and could survive on worlds too hot or cold for humans, and when life got boring they could sleep, dreaming electronic dreams as their spaceships sped across the light-years to the marvels of the next planetary system, where they could build whatever they needed if they decided to stay. Robotic space travelers would never have a housing problem. The elaborate plans for humans to settle on the gravitationally unsuitable Moon and Mars may be the first step in planetary colonization, but that might not last long. If work on artificial intelligence is successful, the first permanent settlers in the solar system and beyond could be well-educated robots. One can see a situation, perhaps only a few centuries away, where humans live in the Earth's biosphere where they belong, and self-repairing and self-reproducing robots go forth to explore those "strange new worlds" that no human crews could ever visit.

ELECTRONIC BRAINS

Of course, we're only mapping our possible future in this way to find a reasonable scenario for what may have happened on other worlds. Perhaps hibernation or deep-freeze techniques could keep biological metabolisms from dying during a few centuries of interstellar spaceflight. But these biological options do not look promising. If a common future for biological brains is to develop computers and then robots with increasingly more efficient electronic brains, then our brightest cosmic neighbors could be robots. The prospect of talking to intelligent robots could be decidedly different from the meetings with biological aliens that we see in films. For instance, the idea of the individual might be partly lost. An advanced civilization might end up with single entities of great intelligence into which many robotic units feed their experiences and information, so that one mind would be at the center of everything. I mentioned this idea in a book published in 1989—before *Star Trek* encountered the "Borg collective," sci-fi's part-biological part-machine development of a nightmare. The problem for the Borg is that the biological part would still need life support and would therefore be vulnerable to all the dangers to flesh and blood. Not a good solution for a space-traveling community.

When we look at how the best animal brain came into existence on Earth, we can see how artificial intelligence could soon surpass our biological intelligence. Since the times of *Homo habilis*, the first human species living in Africa two million years ago, the human brain has increased threefold. The evolutionary pressures and opportunities of a hard and dangerous life caused the human brain to evolve faster than any other major organ in the history of life, as better brains and more dexterous hands provided a better chance of survival for a vulnerable creature. But what made our direct ancestors the first and only technologically intelligent animals on the planet was their ability to simulate detailed actions in the mind before they were carried out. They could envisage tools and weapons for future use before they made them. On the evolutionary timescale, it was a rapid advance, but it nevertheless took a couple of million years. Compare this with what could take place in artificial intelligence. Considerably more intelligence could be built directly into the next generation. We do this already with computers, but eventually the robots might do it for themselves. No need to wait a million years for natural selection to favor the genes that increase mental powers. According to some computer scientists, artificial intelligence may advance to a human level within decades. That could be a takeoff point for an unimaginable future—possibly a dangerous future. Experts of course tend to be too optimistic, but given a couple of centuries, the robots might be traveling to the stars. And what our civilization might do in the near future, those who have preceded us in the galaxy might have achieved at any time since four billion years ago when the oldest sunlike stars may have warmed the first civilizations in the galaxy.

Some biologists believe that no artificial intelligence will ever equal the human brain in its wonderful structural and chemical complexities. That may be correct. The formidable complexity of the human brain is the product of a long evolutionary journey that began with the ancestors of the first primitive fishes several hundred million years ago. Given time, evolution produces great biological complexity, each stage in this process being entirely dependent on what has evolved before. But advanced artificial intelligence will develop in a different way. The approach will be direct and as simple as possible to produce systems that work. At present, no one knows where this technology will lead, driven forward as it is by powerful economic, social, political, and scientific pressures. Already it's an integral part of our civilization, as our reliance on it continues to grow. No one is going to stop this. Take the Internet as one example: no one is going to stop the Internet from further growth and development.

I should emphasize again that all this speculation about our own future is only to try to see the route otherworld civilizations might have followed during the past few billion years. This exercise may help us guess what evidence of our galactic neighbors might be possible. And judging from UFO data, the wise robots from Galactica Incorporated might be keeping an eye on us. Personally, I'd sooner have robots watching over me than weird biological entities. Trying to communicate with super-brainy octopoids could be rather stressful. It's a situation the sci-fi adventurers on television seldom have to face. They usually find articulate humanoids in residence on other worlds and never have to worry about the level of their own IQs when talking to the advanced inhabitants—much like the world explorers from our own history who met isolated tribes and managed to communicate with gestures and odd words. But this was only possible because everyone was of the same species with similar needs. According to UFO contactees and abductees, visiting aliens behave the same way—a sure sign of delusion, which may result from people watching too many sci-fi sagas featuring chatty aliens.

So come back to Earth and see how difficult intelligent communications can be with other species. The chimpanzee is our closest living relative out of several thousand mammalian species, yet we can't talk with a chimp at the same level that contactees claim to have talked with visiting aliens. Chimps can understand human language to a very limited extent, as can other animals, but they can't respond verbally. They can't make the sounds of human speech because they lack the neural and anatomical equipment to do so. Therefore, if we can't really talk with any species on Earth except our own, it's not likely we could talk to beings from distant worlds or that they could talk to us. A basic exchange of information might be possible, but that's not the same as cozy chats onboard flying saucers.

Sometimes, of course, there are no conversations. Instead, the aliens transmit their instructions and comments by mental telepathy—a handy way to avoid the language barrier. Many people in the UFO associations are ready to believe that this occurs. However, even with telepathy, the contactees never learn anything from the aliens that they couldn't figure out from human brains in the bar down the street. Not only is the presence of aliens in chatty mode yet to be confirmed; no one has yet proved the existence of mental telepathy, even between human beings. Bring alien brains into the act, and telepathy seems even less likely, if that were possible.

If mental telepathy is possible, it should have evolved, because it would have had great survival value for intelligent animal life. Evolution has had millions of years to develop such a useful attribute, but it hasn't done so. Since pre-dinosaur days, when the ecological system of predator and prey was well established, a capacity for mental telepathy would have been a great asset. And the case for the evolution of mental telepathy in humankind is even stronger. Experimental psychologists and neurobiologists would love to demonstrate its reality—fame and a Nobel Prize would quickly follow. Actually, they wouldn't have to wait for the Nobel Prize money. They could visit the Australian Skeptics website and claim eighty thousand Australian dollars immediately. Just prove that mental telepathy, extrasensory perception, or telekinesis (for example, spoon bending) exists, and the money is yours. The cash has been waiting for someone since 1980. Perhaps the original spoon bender, Uri Geller, could get together with scientists to relieve the Australian Skeptics of their money?

But apart from mental telepathy, there's an anatomical obstacle to make chatting with aliens difficult. No alien would have a human voice box, a uniquely evolved anatomical structure that makes talking and civilization possible and that is also a very recent product of evolution. Anthropologists tell us that the modern configuration of the human vocal tract does not appear in the fossil record before about half a million years ago, although its evolution to that point stretches back several million years. This precise pattern of evolution is not going to happen elsewhere. Neither is the evolution of the past few hundred million years that started with primitive fishes and the first real backbone, which then evolved to support all amphibians, retiles, mammals, birds, and eventually our ancestors, the first upright apes. The backbone was therefore one of nature's great inventions, and sci-fi films never send anyone into space without one. But could there be similar developments on other worlds? Perhaps so. It's hard to imagine a species building a civilization without the equivalent of backbone, though this is probably underestimating the versatility of evolution.

NO FOOD FOR ALIENS

Of course, robots could communicate in a telepathic way, as my computer and its router communicate wirelessly across empty space in my office. This is yet another advantage robots would have in exploring other worlds, but it's one that the makers of sci-fi films, and some space scientists, still fail to recognize. They see the future exploration and colonization of distant Earthlike planets being performed by flesh-and-blood astronauts. All seem unaware of a fundamental problem: the incompatibility of planetary biospheres that would mean no food for visiting biological aliens. Evolution may create life based on similar chemistries to ours. That seems probable. But those chemistries—the detailed molecular machinery of life—won't be identical. A long series of chance events on a planet and the processes of evolution working on those chance events will see to that.

All biological aliens would presumably eat food and, like us, would use that food to synthesize proteins and other essential molecules to repair their bodies and maintain their living processes. But the essential molecules in their food would not be the same as what we have in our food, since they would have evolved differently. The proteins, amino acids, and all the other molecular components that we use exist in plants and animals on Earth, and we eat those plants and animals to get the molecular components we need to stay alive. But those molecular components have evolved in life on Earth in our biosphere, so you would either starve or be poisoned on another world.

However, in *Star Trek* (the brand leader in interplanetary socializing), we find crews visiting planets for vacations and to collect fresh food supplies. The *Star Trek* writers may know this is chauvinistic nonsense—after all, they are producing sci-fi—but some scientists have actually written about *Homo sapiens* living in other planetary systems, which they could not do unless they had a supply of plants and animals from our own biosphere. The only alternative for biological beings would be to constantly synthesize all food from basic inorganic matter, and that might be too much of a chore, even if it were possible.

It's odd that this problem is not appreciated. A professor of astronomy, unduly concerned with the risk of replying to messages detected by SETI scientists, once said that the aliens might view us "as the finest beef animals they could imagine." This semiserious comment ignores the evolution of proteins and other organic molecules that we find in the "finest beef" and all other food. We know that some proteins can be toxic. There are two dozen snake venoms that are both proteins and potentially lethal, and some spiders can inject poisonous proteins with their bites. Quite enough deadly proteins have evolved on this planet, and proteins that have evolved in the living things of other worlds could be just as dangerous. Much of what has evolved in the cells and tissue of alien life would probably gum up the metabolisms of our best astronauts. And those bloodthirsty Martians that H. G. Wells brought to our planet in *The War of the Worlds*, who ate everyone in sight as they moved toward London, would have killed themselves with protein poisoning long before Wells killed them with bacterial infections and thus saved Earth from Martian domination.

It's nice to know that differences in molecular biology from world to world could prevent visiting aliens from eating us. But could they have other intentions? Another astronomer once suggested that the first ETI (extraterrestrial intelligence) messages would come from "butterfly collectors or missionaries." Butterfly collectors perhaps—and the aliens might take a few samples—but extraterrestrial missionaries? Why would super brains cross interstellar space to convert us . . . and to what? It could be like teaching Christianity to chimpanzees.

CODES FOR LIFE

Some biologists spend their lives studying the genetic code, which determines what we are and how we inherit what we are from our parents and ancestors. In other words, the genetic code (which works at a molecular level within cells) determines the substances and functions of life on Earth. It does this through the production of proteins that form the structures and machinery of life. Because this is so fundamental—with every life-form on Earth using the same code—biologists ask such questions as "Is our code the best possible one?" and "Could better codes have evolved on other worlds?" Actually, their research has shown that there could be many different codes on other worlds, the best codes to evolve being those that would make the least number of mistakes in the production of proteins.

A chemist friend of mine once said that learning the genetic code was a bit like learning the telephone directory. His point was that it's the details that get in the way of understanding, not the concepts. But it's enough to know that our code has to select the right amino acid molecules (from the twenty that life uses) for the machinery within cells to join these molecules together to form proteins. From just these twenty different amino acid molecules, the machinery within cells produces the countless number of different proteins that make life possible And it does this by using combinations of any three of four available molecular "letters" to select the right amino acid molecules, one at a time, that are put in the right sequence to form the protein chain being produced. When completed, the chain folds into the molecular structure that is needed for its particular function.

It's really incredible that this manufacturing process is going on all the time within billions upon billions of cells in every living human body—and within the cells of all other living organisms from bacteria to ourselves. Yet while the information to form proteins is stored in the DNA of every living thing, it is the genetic code that applies this information. No one can say how this complex system evolved, but its existence certainly rests on the universal physics and chemistry as well as on evolution. And as the same physics and chemistry will be available on other worlds, since they are universal, life could form similar systems where conditions allow. However, the way in which our genetic system has evolved on Earth should be unique in its details. No other inhabited planet should have evolved precisely the same system as the one that operates here, even though similar systems may have evolved.

Nowadays biologists talk about the "standard code" because they have found some minor variations but nothing to shake the conclusion that all life on Earth is related genetically. Biologists have also realized that different genetic codes could produce strangely different proteins for life on other worlds. We are assuming here that at least a proportion of life in the universe is based on similar chemistries to the one we use—that "life as we know it" is the norm even though genetic codes are different. We know that the universal physics and chemistry allows this kind of life to form; there are examples of it all around us. But if something like neutron life or plasma life or frozen hydrogen life is possible, then all bets are off. We can only speculate about life as we know it.

It follows, therefore, that on the outside alien life may appear very bizarre while being rather familiar on the inside at a fundamental molecular level. Of course, there may be life that is totally different chemically and physically, but it seems more likely that the ways of putting the chemistry of our universe together to form life that is stable and flawlessly functional may be limited. The fact that on Earth all life is built from the same abundant light elements with traces of certain heavier elements may be the way this universe produces all life.

So it may be that the basic chemical and functional systems of life on Earth are universal, although with lots of differences. Evolution would make that so, but the differences would be variations—perhaps big variations—on the same theme. It may be that no drastically different kind of life is possible, such as silicon life or frozen hydrogen life or plasma life. We see in life on Earth that its outward forms can be as bizarrely different as, say, a human being and the weirdest-looking insect. But on the inside, at a cellular, molecular, and chemical level, they have most things in common because they have evolved within the one isolated biosphere. And so it might be with life on other worlds: great differences on the outside but the same at a molecular and chemical level on the inside. But every biosphere would be different at that level because it would have evolved independently. Consequently, life from one biosphere would be incompatible with life on another, and in reality, those *Star Trek* crews would not find food on any of those "brave new worlds."

The availability of suitable chemicals on a young planet may be the determining factor in how the molecular basis of life on any planet is put together. The first formation of life and its subsequent evolution can only work with what is available. And what has been available on other Earthlike planets may not be the same as what has been available here. Of course, if "life as we do not know it" (something quite different chemically and physically) ever comes our way, we would have no science ready to investigate. Yet even ultra-cold life on Saturn's moon Titan—an unlikely prospect—could in theory use a similar chemistry to life on Earth. What would really flaw us would be visiting robots whose functions were based on science way beyond anything we could understand. And as we've already seen, we can't rule out robotic visitors to the solar system during the past few billion years.

So although messages from aliens might be elusive, the message from biology is clear: that each evolving biosphere will be an incredibly complex system from the molecular level up—and it will be unique. The differences would probably mean that no biological aliens would be able to survive outside their home planet unless they synthesized all their food. Metabolic meltdown could follow the consumption of alien proteins—and would be the fate of any biological alien who tried an Earthly diet. The genetic code here, which synthesizes proteins, would certainly have been as it is today, ever since there were plants and animals big enough to make a meal. This fact alone might have deterred colonizing aliens.

Steven Freeland at Maryland University, a specialist on the genetic code, says, "At the very core of metabolism, we find a surprising number of complex genes (DNA sequences) that are shared by all life on Earth."[1] This points straight to the conclusion that all life here had a common origin. What the "life force" seems to have done on the primordial Earth is to use the most suitable molecules that were available, given that they would do the job, and life on other Earths with different chemical opportunities might have done the same. So the evolved chemical differences in alien life—differences at a molecular level—could be as inevitable as the evolved visible differences of alien life.

MAKING EXOTIC LIFE

Biologists working in this area of science have wondered why all life on Earth uses only four molecular letters to spell out which amino acid molecules are to be assembled into a protein chain under construction. They also specify the sequence in which the amino acids are used. (Biologists call these letters *bases* because they are basic to the structure of DNA and RNA.) So what about experiments to see if life elsewhere could use more letters? This is a big question. Some biologists have artificially increased the number of life's bases (letters) so that amino acid molecules not used by life on Earth could be incorporated into proteins. These are proteins that no life on Earth has ever produced. The exciting aspect of this research is that some of these unnatural proteins could have new and useful properties, hence the great commercial interest in this line of biology.

But we now enter the area of sci-fi speculation about alien life. If life on other planets has evolved to use more than our four letters to spell out its proteins, it could possess proteins whose functions and structures are quite unimaginable to us. Take a feather as a simple example of what I mean. If birds had not evolved feathers, could anyone on Earth have imagined a basic feather, let alone the vast range of different feathers that birds have evolved? It might be a bit like that, though more spectacular on other worlds where genetic codes are larger, offering more scope for the production of proteins with unimaginable qualities. On the other hand, our code might be the optimal one. Four letters directing the manufacture of proteins within cells might be the best number to have. This might be the situation, since we can't deny the remarkable success of our genetic code. However, Steven Benner, a leading contributor to this subject and a professor at the University of Florida, says, "We can't think of any transparent reason that these four bases are used on Earth, and it wouldn't surprise me in the slightest if life on Mars used different letters."[2]

Steven Freeland and Laurance Hurst published a theoretical study that suggests our genetic code is at near optimum efficiency.[3] However, just because biochemists can synthesize different chemical bases in the laboratory and use these to expand our genetic code, it does not follow that such bases would form naturally on other worlds. The subject, though, is still wide open despite the amazing work of molecular biologists in recent years.

A paper from Floyd Romesberg's group at the Scripps Research Institute reports that a number of "unnatural" (laboratory-produced) bases have been identified that rival the performance of the natural bases.[4] So this indicates that elsewhere evolution may have used some of these other bases. We can therefore imagine genetic systems on other worlds with six or more bases supporting vast biospheres of planetary life, which is one more reason for thinking that planetary biospheres may be totally incompatible.

To make a simple analogy, the genetic code can be compared to the evolution of limbs. All land vertebrates have four limbs because the fishes that came ashore some 350 million years ago had already evolved fins that made four limbs possible. Likewise, the six legs of insects and the eight legs of spiders have these numbers of limbs because of their evolutionary roots. For all we know, our counterparts somewhere may have three pairs of limbs; they might be running around on four and using the front pair to build their spaceships. This is a reasonable scenario if they evolved from aquatic creatures that had already evolved the basis for three pairs of limbs.

Therefore, in the chemical evolution of life, as in the large-scale evolution of life, it depends on what has gone before. Alien life might well have genetic systems with six or eight or more bases and might use many more amino acid molecules than the twenty we use to form proteins; some of the biological structures that result might amaze us. Nevertheless, we can't help but notice that a genetic system with four chemical bases has done rather well. Hence the need for samples of life from deep beneath the Martian surface and from the ocean beneath Europa's icy shell to see if different systems have evolved there—unless, of course, some alien biology has come to us. This would be the most convenient way to get to grips with the subject. There's plenty of noise from ufology that alien biology has arrived, but so far there is no substance we can assess for signatures of alien life. Anyone in possession of such a substance could help molecular biologists find these and confirm the major hypothesis in science.

All this shows that there would be a variety of chemical and biological ways to check on the authenticity of visiting aliens—dead or alive. Just a few of their cells would do. Actually, such cells could be more interesting than bodies. In the unlikely event that there were alien cells available, we might find genetic systems that use 6 or 8 or 10 or 12 molecular letters, instead of the 4 we use. Steven Benner's group at the University of Florida has already expanded the number of possible letters from 4 to 12—although in test tube experiments, not in living organisms. And in 2004, the group demonstrated in the laboratory a 6-letter genetic system that reproduced and evolved.

However, no one has yet managed to set up a different genetic system within, say, a microbe, although molecular biologists can make our existing genetic system incorporate an unnatural amino acid molecule into a bacterium's proteins. In other words, the protein-manufacturing machinery of a bacterium (a single-cell organism) can be fooled into using an amino acid molecule that has never before been part of life's proteins. This research, of course, is mainly concerned with understanding the basis of life and with finding new techniques in medicine and in other biotechnologies. But it does also indicate that the genetic codes of life could differ throughout our galaxy and the universe, even if carbon- and water-based life is a universal phenomenon. Therefore, there's no need to argue continuously about the authenticity of any fragment of an alien organism, as some ufologists repeatedly do. Just let the biologists have a few cells. If its protein profile matched that of life on Earth, it could not be an alien. The people who have tried to hoax us with evidence of aliens haven't allowed for protein profiles. And some ufologists, unaware of this relevant biology, even argue for a genetic link between prehistoric humans and visiting aliens. At a lecture some years ago, a man became quite disturbed because I explained why he couldn't have had an alien ancestor. He wanted superior beings in his family tree, not a lot of ancient apes. At least some of our genes, he argued, must have come from advanced aliens. How else could humans be so intelligent?

I believe the phrase "life as we know it" has confused many people. "Life as we know it" does not mean identical life to ours right down to the molecular level, with all the molecular machinery working in the same way. It means life based on the elements and chemistry that make life possible on Earth. Life here has used these elements (principally hydrogen, nitrogen, carbon, oxygen, sulfur, and phosphorus) to make and run the machinery of life with such amazing efficiency that we may speculate that similar chemistries do so throughout the universe where conditions allow. However, because evolution must be a factor in this, even at a molecular level, the precise form the molecular machinery of life has adopted can only vary from biosphere to biosphere.

Science fiction can be dreadfully misleading on this subject. Not long ago, I watched a *Star Trek* episode in which a small alien craft was detected and brought aboard. Inside this craft the crew discovered a solitary occupant who was a very dead humanoid. The doctor analyzed its DNA and found that the dead occupant had genetic sequences from several different humanoid civilizations. In other words, it had its origin in several different alien biologies, all of which, according to the story, would have had to possess genetic codes that matched precisely. Unless the "several different humanoid civilizations" had a common biological origin, this is next to impossible. But no one onboard, including the doctor, had a problem with this.

HERE BY CHANCE

Now, what the man at my lecture wouldn't accept was that the evolution of humans is well supported by mainstream science, including the extensive fossil record and knowledge of the conditions in which our apelike ancestors lived for a few million years on the savannas of Africa. Our arrival on the planet didn't need help from aliens, and nothing in the fossil record would lead biologists to think otherwise. Indeed, the growing fossil record is beginning to show that numerous species of hominids (upright apes) evolved and lived in the same period on the continent of Africa from about five million years ago and that they spread out from there. All eventually became extinct except our own ancestral line, though the Neanderthals survived in arctic conditions until about thirty-five thousand years ago. Had the climate improved at the right time, the Neanderthals might still be with us—given that an expanding population of *Homo sapiens* allowed such a situation to continue.

The 2004 discovery of the fossils of miniature humans on the Island of Flores in Indonesia was a tremendous surprise. These people grew only as tall as a three-year-old child and had a small brain, yet they made tools, used fire, and hunted pygmy elephants and other prey. According to the scientists who studied this little cousin of ours, these miniature humans lived on the island from 95,000 to 13,000 years ago. One theory about their existence is that the original inhabitants were *Homo erectus* that, isolated on the island with limited food resources, evolved a much smaller statue. Given that all this is roughly correct—and the subject is still open—humanoid evolution is a lot more flexible than was previously supposed.

THE TROUBLE WITH HUMANOIDS

Finding worlds with life is expected, even if Mars and Jupiter's moon Europa are sterile. There could be many planets supporting life in the newly discovered planetary systems in our part of the galaxy, though this life should not include different species of humanoids. *Star Trek* might find humanoids on every inhabited planet in the galaxy, but biological reality would be different. So there is a problem when witnesses in close encounters with flying saucers describe the occupants as humanoids. Even when robots are reported, they have a humanoid form. What's going on? Apart from the hoaxers and people prone to fantasy, are the witnesses seeing what they expect to see after years of watching sci-fi films, or could some reports be roughly accurate? Some witnesses have reported "saucernauts" that are 100 percent human; others have described visitors as short and slender or tall and ghost-like. The most common type is the so-called grays, who are small with large heads, tiny mouths and nostrils, and enormous black eyes. The grays get plenty of publicity in the media, but other occupants are also humanoid in their fundamental anatomy.

The differences between these visitors and us are not any greater than, say, the differences between us and our nearest relative, the chimpanzee. The problem with these humanoids is that gray skin, big heads, no hair, pointed ears or no ears are all superficial differences on a fundamental anatomical structure: the humanoid body plan. So could this body plan evolve on other worlds? Because form and function go together in all life-forms, some technological aliens might have evolved a similar body plan to our own on the basis of different biochemistries and internal structures. On the outside, they'd be roughly similar to us, though if we examined their internal organs and the cells of their tissues, we'd find big differences because these would have evolved independently during vast periods of a planet's history. Of course, it would be more acceptable if witnesses consistently reported weird octopoids climbing out of their flying saucers. However, if we accept—provisionally—that some witnesses have seen humanoids, we can slip into sci-fi mode and offer some explanations.

1. The visitors have evolved from our ancestors captured long ago by the real aliens who might be either biological or robotic—or even weird octopoids. This would explain why they can move naturally under our gravity and breathe in our atmosphere. It might also explain their detached attitude toward us, their distant, unstable relatives.

2. The limitations on evolution are much tighter than we think, so that only the humanoid life-form can become a technological life-form. But there's a time-frame problem here: two similar humanoid species evolving independently in our galaxy in more or less the same period. This could be feasible only if the evolution of humanoids is far more widespread than seems possible. Most biologists would say that the evolution of a single species of humanoid anywhere else in our galaxy is unlikely. Alternatively, we could imagine that the only life-form to develop space travel was a race of humanoids long ago who, over millions of years, established themselves in many planetary systems. This cuts down the odds a little because those humanoids would not have evolved at the same time as us but millions of years earlier. Yet we still have to ask, how are they able to visit us in our epoch when the time frame of opportunity has been so vast? This problem is never mentioned in the UFO literature, which seems to accept that a whole range of humanoid species could evolve on different planets at about the same time as we did. We get this same view on sci-fi television, where different alien humanoids are constantly meeting in the vastness of interstellar space. This impossibility may make sci-fi drama possible, but along the time line, which is at least eight billion years long, any new technological civilizations would have emerged at different points. This makes it impossible for two civilizations in our galaxy to be at the same level of development at the same time—a big obstacle to socializing with the aliens.

3. Visiting aliens think their shockingly different appearance would create fear and hostility. So what do they do? They use intelligent robots similar in form to the planet's ruling inhabitants to do the legwork. This sounds like science fiction, and it probably is. But it's less like science fiction than having humanoids evolving everywhere in the galaxy in the same epoch.

4. Some physicists say time travel is possible, so let's have the humanoid visitors coming in their flying saucers from a future Earth, where our present humanoid form has evolved into something more fragile and flimsy and more intelligent.

Take your pick of any of these options. And if anyone has another explanation, send me an e-mail. If anything, I favor the ultimate in robot technology that would need only energy from some source. Not exactly an interesting diet, but highly intelligent robots should not have food on their minds. The character Data on *Star Trek* may be a robot who enjoys his food and drink aboard the starship *Enterprise*, but one can't help wondering what happens to that food and drink. Maybe the *Star Trek* creators will tell us sometime?

A further development that destroys the credibility of reports of humanoid aliens comes from some UFO enthusiasts who are not satisfied with just one species of visiting humanoid. They have them coming from all parts of our galaxy, even from other galaxies, which seems unnecessary when our galaxy could have a few billion planetary systems. Yes, the humanoid body plan is ideal for creating civilizations, as we know, and something like it may have evolved elsewhere. But those multi-limbed octopoids might be saying the same thing about their own body plan.

We haven't got a lot to go on, but we see from life on Earth that there may not be a limitless number of basic body plans in the universe. By "body plan," I mean what biologists call an "animal phylum." It's part of a system of classification. The animal kingdom is divided up into about thirty phyla; each phylum includes all those living systems with the same animal body plan, like insects and vertebrates and spiders. All animals with backbones belong to the vertebrates, although most animal species are invertebrates. Several animals with invertebrate body plans lived long ago, including some very odd creatures, but we only know them from their fossils.

Plants have a different system of classification, but plants need not concern us here. No plant is going anywhere in the evolution of intelligence, though this hasn't stopped sci-fi filmmakers from showing intelligent vegetables attacking our planet. Even humanoid vegetables have figured in sci-fi dramas, though vegetables as we know them have not evolved a single nerve cell, let alone something like a human central nervous system of a hundred thousand million nerve cells. Nerve cells need a lot of energy, and plants just can't supply it. But they don't need to. They've been a great success without nerve cells.

At this point, I would like to speak a little heresy. If we start by noting that more than thirty successful animal body plans have evolved on Earth from, say, the level of jellyfish upward in complexity, we might ask why this number is so limited. Why aren't there hundreds of different body plans? The answer must be that since the explosion in animal evolution some six hundred million years ago, life hasn't been able to evolve them. This means that the combination of Earthly environments and life's genetic machinery can't extend the range of viable body plans. So we have an intriguing question here. We've seen that only one body plan, the vertebrates, has evolved large brains. No animals in any other phyla have evolved even moderately sized brains, except for some cephalopods (squids and octopuses), and some animals don't have brains of any sort. Moreover, only one line of evolution, the primates within the vertebrate group, has led to a species with technological intelligence. Therefore, despite the vague technological potential of the octopoids, which has never been realized, it might be that only life-forms somewhat like vertebrates would produce creatures capable of technology, and that those highly intelligent monsters in numerous movies have no place in the real universe.

So it's not impossible that if we were being visited by real aliens we might recognize something like our own body form. This, of course, is what some accounts of UFO occupants provide, although none offer any acceptable evidence. Alternatively, some cataloged reports from pilots and aircrews of mysterious lights and strange aerial craft do provide a semblance of credibility. Reports of odd-looking humanoids badly need something to give them credibility. Just a few cells from a "ufonaut" body would do nicely, though of course a few bodies would be better. They might show the limitations of evolution in its production of technological creatures. So if anyone has a dead alien on ice from a saucer crash, there's no need to produce the whole body—not yet. Just send a few cells along to your local university so that biologists can begin to test for an alien signature of life. This would be the easiest way to confirm or repudiate the existence of dead alien. The biologists could do it, but can ufologists provide the cells?

BOGUS BODIES

Although no visiting aliens have left their bodies behind—judging from the lack of evidence—we have seen a few in dubious documentary films. I watched a film featuring alien bodies some years ago that pretended to show aliens being examined by doctors after the legendary saucer crash near Roswell in the New Mexico desert in 1947. According to the story, an unidentified cameraman taking photographs of the crash site for the US military kept some footage for himself. After all, it's not every day you get to photograph aliens. Eventually, he sold the film to an enterprising British film producer. The media was quite carried away by the story and spread news of it around the world, although, to my knowledge, no one bothered to check with the science community.

I saw just one dead "alien" in the film. Every bone, every joint, every muscle was identical to the species *Homo sapiens*. The hoaxers—having received their biological education from sci-fi on television—assumed as usual that all visiting aliens would have their bones and muscles in the "right" places. Even today, some UFO websites still support the authenticity of those filmed bodies. One quotes an unnamed pathologist who saw the film as saying, "Although a close-up of the brain was shown . . . the appearance was not that of a human brain." Another pathologist, again unnamed, is reported to have said, "As for the organs removed, they could not be tallied with any human organs."

What an amazing phenomenon! On the outside, the aliens are anatomically identical to human beings, but on the inside, they're radically different. A real pathologist would spot such an inconsistency. The lesson for hoaxers is clear: if you want to bamboozle us with films of aliens, dead or alive, you should study biology. Even a weird alien body would not necessarily be acceptable. Hoaxers think they've done a good job when they display what looks like an anatomically "correct" body, but they never think about an alien's protein profile—which would actually be more convincing than the body itself. Even the biochemistry and molecular biology of some genuine alien microbes could provide a revelation in our understanding of life—we wouldn't need a whole alien body. We would be able to see in alien microbes how alien biochemistry and molecular structures worked and where they were different from those of life on Earth. There might be similarities, or things might be very different.

We can't just guess what would be discovered, and that's why the claims from ufologists that we've had access to dead alien bodies—or live alien bodies— are unbelievable. At present, for instance, we don't know where the limitations of physics and chemistry end and where the mechanism of evolution begins. (Keep in mind here that the fundamentals of physics and chemistry will apply everywhere.)

The chemistry of life shared by all life here is so staggeringly complex that many processes cannot yet be explained in evolutionary terms. We just don't know enough, though one day it should be possible. It's really amazing that any life is fully functional, since the complex chemical systems essential to life can be made inoperable by the removal of just one step in a chain of metabolic processes. And just one faulty gene can make life intolerable or impossible. So we can't help wondering how evolution has produced our amazing systems of life, though it has had a few billion years to do this. Perhaps, for optimal function, those systems had to evolve more or less in the ways in which they have done so here. Maybe the biochemical options for evolution were rather limited. If so, we'd find similar evolutionary results at a molecular level on other worlds.

Many excellent books on evolution pass over the staggeringly complex biochemistry that makes life possible and give the impression that all aspects of the evolution of life have been worked out. This is not so. However, a wide range of scientific disciples confirms Darwinian evolution, and no biologist doubts its validity. The evidence for natural selection as the engine of organic change is overwhelming and is probably a hundred times greater today than when Darwin published *The Origin of Species*. For this we have to thank thousands of scientists working in paleontology, geology, physiology, anatomy, microbiology, genetics, molecular evolution, and biochemistry—to name some of the relevant disciplines. Evolutionary theory may not yet be able to explain in terms of natural selection all the incredible complex chemistry and molecular structures of life, but it is the best-supported theory in science. And if biologists could prove Darwin wrong, they would do so. An illustrious place in the history of science would await them. I tell this to the Jehovah's Witnesses who knock on my door and who want me to replace The Origin of Species with the Bible. I tell them they should read both, but I doubt if any do.

THE ROSWELL LEGEND

Published in 1997, *The Day after Roswell* is a good example of how some prominent UFO books fail to deliver what scientists expect, if a subject is to be taken seriously.[5] The book is supposed to be the work of a distinguished US military man, the late Colonel Philip Corso, but another name appears in small print on the cover—that of the ghostwriter (it seems likely that the ghostwriter's imagination and input overwhelmed the colonel.) Any sci-fi writer could have provided a description of the examination of dead aliens from the legendary Roswell saucer crash. If biologists actually had access to real alien bodies, their research reports would be full of surprising but believable details. A scientific presentation of the details of alien life, from its biochemistry to its anatomy, would have the stamp of authenticity. What seems inexplicable is that the colonel was reviewing for his superiors the technical reports from scientists, yet there is no scientific detail whatsoever in what he has to say.

I'll add that it's not only on the biological front that the colonel's account (or that of his ghostwriter) doesn't ring true. According to the book, the major technological advances in the second half of the twentieth century were all due to the United States back-engineering the advanced alien technologies found in the crashed saucer. I quote: "Today, items such as lasers, integrated circuitry, fiber-optics networks, accelerated particle-beam devices, and even the Kevlar material in bulletproof vests are all common-place. Yet the seeds for the development of all of them were found in the crash of the alien craft at Roswell."[6] The numerous scientists whose hard work and intelligence led to these technological developments could be a bit upset by being told that credit should go to the aliens. But anyone can check the facts for themselves. The histories of all the new technologies mentioned by Colonel Corso are well known. Some of the scientists responsible received Nobel Prizes for their work. Was the colonel (or his ghostwriter) saying that the scientists received prize money and acclaim that should have gone to aliens? Anyone who thinks there could be a fragment of truth in this kind of back-engineering story needs to consider just one thing: technology is based on science, and new discoveries in science make new technologies possible.

If you don't understand the alien science, you're not going to be able to back-engineer the alien technologies.

After reading Corso's book, I found a quote from Webb Hubbell, who headed the Clinton administration's Department of Justice. According to Hubbell's book *Friends in High Places*, the president told him, "If I put you over at Justice, I want you to find the answers to two questions: One, who killed JFK? Two, are there UFOs?"[7] Hubble was never able to provide answers. If Clinton couldn't get an answer to the UFO question and the claims about the Roswell saucer crash, who could?

Well, a lot of people on the web claim to know the answer. These days you don't need to travel to other planets to find your cosmic neighbors; the websites will bring them to you. They not only support the authenticity of those mythological alien bodies from the Roswell crash; some also provide detailed classifications of numerous alien species currently visiting Earth from different parts of the galaxy. They're all humanoids, of course, and their characteristics are well presented, well described—and a lot of baloney.

On one site, the aliens' credentials are so well presented that it must be a hoax, perhaps to deter scientific interest in the subject of alien life. The site lists sixteen different species of alien visitors, all humanoids who, quite amazingly, have evolved in more or less the same epoch in different parts of our galaxy. Darwin would be astonished, as would all present-day evolutionary biologists. The reptilian humanoids, we are told, have genetics akin to our reptiles, although the identities of the geneticists and molecular biologists who must have labored long and hard to obtain this information are not given. Neither is it mentioned that our own genetics includes genes from early retiles—before the dinosaurs—that gave rise to the first mammals. We even have genes from fish that died a few hundred million years ago. When we were embryos, we all had gills that grew from instructions in genes inherited from our fishy ancestors.

That imaginative website also tell us about the "Gray Type A," a humanoid type of alien that likes conquering worlds—though we're left wondering why they waited so long to conquer the Earth, given that it has been here, unconquered, for a few billion years. These particular grays have "insectoidal genetics," and they've been "trying to cross breed with humans," which may take some doing, considering that insects, unlike reptiles, are not part of our ancestry.

Other visitors include tall, blond aliens; goblin-like aliens; and hairy humanoids, but I'll let readers check this out for themselves. From the boldness of their classification, you'd think the people who manage the website had a wide range of alien species in their deep freeze and an army of biologists to study them. It's obviously a hoax, but some people may believe it—that's the problem.

Of course, you don't need visiting humanoids to confirm that life is universal. Searches for life within the solar system (under the surface of Mars or in water beneath Europa's icy shell) may find little more than microbes, but alien microbes could revolutionize our view of life, providing their origin was completely independent from life on Earth—something that could be determined from their molecular biology and genetics. But, as we've seen, there's a problem. Life on Mars, if it exists, could be the result of splash-offs when large meteors hit Earth, just as rocks from Mars have been found in the snows of Antarctica. One can also imagine rocks containing microbes from Earth or from Mars crashing into the ice crust of Europa, although that's a long way for bacteria to travel encased in rocks, and Europa's ice may be many miles thick. So life in the warm internal sea of Europa—if there is any life there—is more likely to have had an independent origin. But we could be Martians. It's possible. Life could have started on the Red Planet. But in that case, our Martian ancestors would have had to arrive here about four billion years ago. The fossil record indicates that life got started here at about that time.

PATHS TO TECHNOLOGY

We can find the uniqueness of *Homo sapiens* confirmed in the Human Genome Project, on which molecular biologists and geneticists work to relate all human genes to their functions. The scientists involved have calculated that the number of genetic letters used to spell out the recipe for a human being is equal to the number of letters in thirteen complete sets of the old *Encyclopaedia Britannica*. But it's not the large number alone that makes the evolution of a recipe for a human being elsewhere in the universe impossible. It is the sequences of the genetic letters that spell out a human being. Like the letters in a book, the molecular letters have to exist in a certain order. Now, the probability of another biological evolution bringing together millions of genetic letters in the right order to produce a human being is going to be nil in the whole of space and time in a finite universe. The only way out of this corner would be to have an infinite universe, or a universe with an infinite past and an infinite future, or an infinity of universes. But that's getting too cosmological and metaphysical for this book.

What is relevant here is the way organic evolution works as an integral part of creation, as much an aspect of our universe as, say, the inverse square law or even the three dimensions within which we have our being. Organic evolution would automatically come into operation anywhere in the universe with the origin of life (with the emergence of self-replicating systems), but it took four billion years on Earth to advance from the simplest cells to creatures with big brains. It might have taken more or less time elsewhere, but the point for SETI, which must not be forgotten, is that organic evolution on other worlds could have begun in our galaxy some eight billion years ago, given the ages of some sunlike stars. That in theory gives plenty of time for the early aliens to have discovered all planetary systems with life. Few biologists and astronomers would argue against this possible scenario, but it's only a scenario. The evolution of space-age beings demands a lot more than the evolution of complex life-forms. We have only to consider our own evolutionary path to appreciate this.

Before creatures capable of creating civilization could evolve, there was a long series of evolutionary events from the marine ancestors of the first vertebrates some 450 million years ago to the past few million years when the first upright apes appeared. From what has happened here, the evolution of a primitive backbone could be viewed as a beginning on the path to civilization, but the fishes that carried this feature forward might not have been as successful as they were. There was competition from nonvertebrates at the time, yet a species of fish eventually took the backbone onto dry land for the evolution of the first amphibians, which led to reptiles, mammals, and birds. But the most recent major evolutionary advance took place after dense forests were replaced by dangerous open savannas. This put evolutionary pressure on numerous species of upright apes to become more able and intelligent. And this happened at the right time in Earth history to create a path from life in the trees to life in cities. One wonders what would happen on a world with no counterparts to our vertebrates and monkeys and apes? Life's path to civilization, if such a path has ever been followed before, might have been amazingly different.

We might wonder, if all *Homo sapiens* were removed from this planet, would any of the present species of apes lead to a technological species—or would some other mammal make the grade in another fifty million years or so? We can't answer such questions. We can't even put a figure on how likely it is that a technological intelligence would evolve in a flourishing biosphere. And we have no idea what the average period would be for that to take place. There were times on our own evolutionary path to big-brained vertebrates when the process might have been quicker, but there were also times when it might have been slower or might not have happened at all. For example, the dominance of the dinosaurs for 160 million years—until 65 million years ago—kept the mammals from evolving larger animals with larger brains.

It's a conspicuous fact that all big-brained animals on Earth are both vertebrates and mammals that have metabolisms to maintain constant internal conditions for life's biochemistry. The only exception is the cephalopods (octopuses and squids), which are the most intelligent animals without backbones. But the cephalopods first evolved five hundred million years ago in the Cambrian seas, and many fascinating species of cephalopods still live in our seas, showing no sign whatever of evolving bigger brains. So might most otherworld civilizations be established by technological creatures whose ancestors, like ours, could only initially evolve in an aquatic environment but who later colonized dry land? We might also ask if the backbone, on which everything can be attached, is a common structure for land life and the evolution of animals with large brains. There's great variation in backbones on Earth, of course, but they're all related, evolving from a prototype in the primitive fish that lived around 450 million years ago. And because most advanced animals from that time on had a front end and a back end, a supporting structure connecting the two "ends" seems a likely development, wherever in the universe animal life might have an existence. However, all alien backbones, if there are any, would have a different prototype and different evolutionary pathways. And we'd soon see that they weren't related to any backbone on Earth.

UNKNOWN LIFE-FORMS

As we discover more about life on Earth, the probability of life elsewhere increases. We now know that microbial life thrives in hot springs, glaciers, at miles below ground, and in other extreme environments. What we call "extreme environments" are considered home to the microbes, and how they manage to make a living in such environments is a research subject that is relevant to searches for life beyond the Earth.

Until the revolution in molecular biology during the past fifty years, the classification of life had not changed much since the days of Darwin. There were five divisions: the plant kingdom, the animal kingdom, the fungi, single-celled life like the amoeba (some of which could have the characteristics of both plants and animals), and the bacteria. The bacteria were quite different from the amoeba level of single-celled life, having far less complexity within the cell and being much smaller. So the bacteria were regarded as the smallest and simplest life-form; the viruses were not thought of as "life" because they need the cellular machinery of host organisms in order to reproduce.

But life on this planet is always showing itself to be more complicated than we suppose, and not long ago, the accepted classification of life got a fundamental revision. In recent years, the molecular biologists and geneticists have enabled us to view the divisions of life in terms of their evolutionary histories, rather than by their outward appearances. In other words, we determine their histories and relationships by their DNA sequences (their evolved genetic programs) and what these produce. In this way, we can see relationships that could never have been seen from the outward appearances of living things.

And so it was in the 1970s that biologists began to discover what they thought were bacteria in some of the most hostile environments on Earth. These microbes were living in deep-ocean volcanic vents in water that seemed far too hot for life and where energy from the sun never reached. Soon after this discovery, oil companies began to bring microbes up from several miles beneath the Earth's surface where they were nourishing themselves chemically from surrounding substances. To have living microbes from such extreme environments was a great surprise in itself. It increased the extent of the biosphere. But there was a much bigger surprise to come.

When molecular biologists studied the genetics of these "bacteria," they found that the majority of microbes in extreme environments were different from bacteria and quite unrelated genetically. The sequences of their DNA showed that they had no more in common with bacteria than they had with plants and animals. They were so different that their DNA sequences showed that they must have separated from real bacteria almost four billion years ago, and that since then, they had followed their own evolutionary pathways. These microbes were initially named *archaebacteria*, the initial thinking being that they actually were archaic bacteria. But after the DNA sequences showed them to be distinctly separate from bacteria, they were renamed *archaea*. Consequently, and based on modern genetics, we now have three domains of life on Earth instead of five. They are the bacteria, the archaea, and the rest. And the "rest" includes plants and animals, the fungi, and all those advanced single-celled life-forms like the well-known amoeba.

But there was a further surprise. It was found that not all archaea are extremophiles, thriving in near-boiling water or living deep beneath the Earth's surface. About twenty-five years ago, marine microbiologists discovered a wonder bug in the Sargasso Sea that has turned out to be the most abundant life-form on Earth—and it's an archaean. Its name is *Pelagibacter ubique*. Dr. Hazel Barton of Northern Kentucky University, one of the specialist microbiologists working on *Pelagibacter* says, "It composes about 26 percent of all the micro-organisms that live in the seas and oceans of the world. It lives on nutrients that we can't even see, in water that is cleaner than distilled water that you might put in the battery of your car."[8]

Obviously, microorganisms that have evolved metabolic systems to live in such conditions are very abundant because vast regions of our planet offer such conditions where other life-forms can't make a living. Since its discovery, *Pelagibacter* has been found in other oceans where the water contains practically no nutrients. How it manages to obtain enough energy to be such a successful microbe is a problem for microbiologists, but such discoveries lead us to think that life—at least in the microbial form—could be very common in the universe.

Professor Steven Giovannoni, a marine microbiologist at Oregon State University, says, "Diverse communities of Bacteria, Archaea and Protista (the advanced single-celled organisms such as the amoeba) account for more than 98 percent of the ocean's biomass. These microscopic factories are the essential drivers for all of the chemical reactions within the bio-geo chemical cycles on Earth."[9]

The fact that microorganisms constitute the major part of Earth's biomass (life by weight) may seem contrary to our commonsense view, but it's a fact. Microbes are everywhere, and plants and animals are only the visible peak of life supported by this abundant and ubiquitous microbial realm. Earth's biosphere could not otherwise exist. You could take away all the plants, animals, and fungi, and life would continue to flourish in a microbial form. But take away all the microbes, and the rest of life would die.

I remember a group of biologists in the 1960s, when the space age was booming, trying to breed germ-free chickens. The crazy idea that prompted this research was that future astronauts might like fried chicken on the menu while traveling to Mars, so why not take live chickens with them? But the germ-free chickens all died because they couldn't live without their microbes. All animals would suffer the same fate.

So, is microbial life, which has flourished here for more than 3.5 billion years, relevant to SETI and our view about life on other worlds? Well, what has taken place here could have taken place in roughly similar ways on planets like Earth. And where advanced life exists, microbial life-forms will have flourished to provide the basis for it, as well as maintaining the chemical stability of any planetary biosphere. This means that any visiting aliens who aren't nonbiological robots are going to bring their own microbes with them.

A few years ago, I saw a television show that played a tape of a small humanoid alien sitting at a table, supposedly at a secret US government laboratory. A man was interviewed about the tape, which he claimed to have smuggled from the laboratory. The alien coughed a lot and seemed unwell. The television interviewer asked about its health and whether people nearby might catch some troublesome alien disease. The interviewee quickly assured him that they couldn't, that the aliens had long ago rid themselves of all microbes and infections. At this point, I knew I was watching a hoax. All life-forms on Earth, including ourselves, are ecosystems maintained by microbial life, and the same state should exist on other inhabited worlds.

All this indicates that the most likely form of life to exist on other habitable worlds would be microbial life. And it could be the first alien life to be discovered, perhaps on Mars, unless we receive a broadcast from the stars or some aliens get caught on camera with their flying saucer. But we can't rely on the aliens confirming the Grand Hypothesis in this convenient way, so in the next few decades several groups of astronomers, using space telescopes, will be trying to do so. They will be studying the newly discovered planetary systems for certain spectral lines, signatures of life that indicate its presence. And if they are successful, the SETI scientists are sure to start scanning these newly discovered worlds for intelligent signals. However, from what we know of life here, those life-marking spectral lines are more likely to be produced by vast populations of microbes than by vast populations of clever aliens. And where very intelligent life-forms haven't evolved, there will be no broadcasts.

Judging from Earth history, photosynthesizing microbes would take a long time to make a planet fit for creatures that could create civilizations. In our case, microbes and plants eventually raised the oxygen level sufficiently to form an ozone layer, which protected life as it left the radiation protection of water and colonize dry land. This could be a common process on Earthlike planets. Astronomers might confirm the presence of life on other worlds by detecting the prominent spectral line of ozone (formed from three oxygen atoms) and other biomarkers, although this is a long way from the main goal of traditional SETI, which is to discover our counterparts in the galaxy.

There's another complication to this business of finding anyone at home who's more advanced than a bacterium. The fossil record shows that advanced life wasn't established here until some 550 million years ago, when most of today's life-forms appeared, although the vertebrates—those animals with backbones and the best brains—do not appear in the fossil record for another hundred million years. It has always been a puzzle why microbes did not lead to more complex life much earlier. Yes, there had to be enough oxygen in the oceans and the atmosphere to support larger life-forms, and an ozone layer had to be in place to protect the evolution of land life from radiation. But research shows that an adequate level of oxygen and the ozone layer existed a long time before life colonized dry land. We can't help wondering if the vast time period between flourishing microbes and active intelligent life-forms might have been a lot shorter than it was on Earth.

Perhaps the concept of a "snowball Earth" is relevant to our discussion. In recent years, geologists have found evidence that the Earth was completely covered in ice and snow at least once in its history. Ice may also have covered the oceans, although geologists are uncertain about that. But rocks from various parts of the Earth do bear the recognizable scars from flowing glaciers, and some of these rocks would have been at the equator. Although microbes could have survived in the oceans beneath the ice, the Earth during those periods would not have provided encouragement for anything much more advanced to evolve.

So, massive ice ages might have delayed the evolution of intelligent animals. And earthlike planets that escaped being "snowballed" might have supported civilizations a billion years before us, even though their stars are about the same age as the Sun. Had such alien civilizations arrived here even a billion years ago, they would have found nothing but microbes—and sometimes, perhaps, a snowballed planet. Yet the molecular machinery of native microbes would have been interesting, given that the chemistry of life in the universe has followed a broad path with plenty of options for evolution to exploit. Just how broad the path is and by how much evolution is constrained in what it can produce might have been something any alien visitors would have been keen to study. They would have taken samples and entered details on how the newly discovered life was running its basic chemical processes, and then left for more comfortable and interesting planets.

THEOLOGIANS, EVOLUTION, AND THE MULTIVERSE

It may not be immediately obvious that the science of SETI explores deep regions of knowledge that are relevant to theology. Most of the theologians I've met over the years are worried about evolution on Earth, especially the uncomfortable fact that our ancestors were a lot of enterprising apes. But those apes are not really significant for theologians. When Darwin published *On the Origin of Species* in 1859, Victorian theologians were not ready to see the significance of evolution as an inevitable part of the way the universe works. They didn't see that the process of natural selection would inevitably follow once assemblies of molecules began to reproduce themselves. The successful groups went on to reproduce; the unsuccessful did not. And any chance changes that improved this process of reproduction would have been kept.

So the process is fundamental. Natural selection will take place anywhere in the universe where entities reproduce while competing for survival in life-supporting environments. However, the complex structures and processes of life at a molecular level can be determined only in part by natural selection. The existence of every aspect of life rests on the universal constants of nature (the universal physics and chemistry) being precisely as they are so that the molecular structures that support life are possible. This means that the atomic elements used by life (twenty-one of them) have to possess the necessary characteristics to form the molecules that fit together and run the machinery of life. Any theologians who don't like the look of their apelike ancestors should focus on these aspects of the universe, because this is where theological bedrock lies. The fundamental constants, plus the nature of the available chemistry of the atomic elements used by life, are not the products of Darwinian evolution. They do set the limits for what is possible for life, but we don't yet know to what extent the universal physics and chemistry determines what is possible or how many ways there might be to produce the phenomenon of life. Only knowledge of life on other worlds could begin to tell us about that. But some consideration of the fundamental constants and the chemistry of life is very relevant to all aspects of SETI—and theology.

THE FUNDAMENTAL CONSTANTS

Let's start at the beginning of things and work forward in time and upward in complexity. That means we start with the fundamental constants, a remarkable series of relationships between matter and energy that govern all aspects of life and the universe. The degree of fine-tuning is amazingly precise, and no one can say how the fundamental constants happen to have the values they possess. It's a great mystery. When challenged by theologically minded people who worry about evolution, I sometimes point out that they should consider the fundamental constants that lie at the deepest level of our awareness of the universe—where we also find support for the hypothesis that life and intelligence are universal. Some people may think it strange that SETI could have any conceivable link with theology, but if any of the fundamental constants had a different value, we wouldn't be here to search for other life, and there wouldn't be any other life to search for. But because the fundamental constants are as they are, the origin and evolution of life could take place anywhere in the universe where conditions allow it to do so. The fundamental constants cannot apply to us alone, which means that, with billions of planetary systems just in our galaxy, SETI should have plenty of alien life to discover.

So how different would the universe be if the fundamental constants were different? This would depend on which ones were different and by how much, but for the most part, life of any kind would be impossible. Astronomers often say that we live in a hostile universe, and in one way this is true. Astronomical photographs show the results of colossal stellar explosions that rip across galaxies, destroying everything they meet. But at its most fundamental level, the universe is amazingly benign toward life. As the great astronomer Fred Hoyle used to say, "It all looks like a put-up job."

Let's look at some of the fundamental constants, just to give us a feel for their significance for SETI.

(a) We'll start with the charge on the proton, which is equal and opposite to the charge on the electron, even though the proton has 1,836 times more mass than the electron. Physicists tell us that they have measured this equivalence of the charge to an accuracy of one part in a million billion billion. A slight change in this equivalence (of the proton's charge to the electron's charge), and atoms would never have formed in the first place.

(b) The primary energy source in the universe is the release of energy when hydrogen is "burned" in stars to form helium, the next-heaviest element, converting about 0.7 percent of matter into energy. The energy comes from the disruption of the so-called *strong force* that binds nuclear particles (protons and neutrons) together. Astrophysicists say that if this force had been just 1 percent stronger, the universe would have had no free protons in its early days to form hydrogen atoms. Consequently, the universe could not have developed because hydrogen is by far the most abundant element in the universe and provides the basis for the synthesis of all heavier elements. Without hydrogen, there could be no galaxies, no stars, no water, and no life. We are here, the phenomenon of life exists because the strong force, the binding force is precisely as it is.

(c) If we go the other way—if the strong force (binding force) had been 1 percent weaker, then the synthesis of helium from hydrogen (nuclear fusion in stars) would have been impossible. The reason? A stage in the synthesis of helium involves making deuterium, a heavy form of hydrogen with an additional neutron loosely bound to its proton. With a weaker binding force, deuterium could not have held onto an extra neutron and would have been too unstable to support the synthesis of helium. Hydrogen fusion into helium would then not have taken place; the stars would never have shone, and no heavier elements could have been synthesized.

(d) Gravity is nature's weakest force, though you don't much appreciate the fact if you fall off a ladder. It's a hundred thousand billion billion billion times weaker than the electromagnetic force, which keeps electrons "orbiting" their atomic nuclei. But move from atomic nuclei to the far greater mass of the Earth, and gravity becomes a controlling force. Astrophysicists say that if gravity had been much weaker, the uneven distribution of matter in the early universe would not have been pulled together to form galaxies. So, no galaxies, no stars—and no energy for life.

However, had gravity been significantly stronger, the matter forming the galaxies would have been pulled together too tightly. Stars would have formed closer together, smaller and denser than the stars we know, and they would have rapidly "burned" their nuclear fuel. Stars like the Sun, in a denser form, would have consumed their energy resources thousands of times more quickly than such stars do in our universe. No suitable planets for life could have orbited such stars.

(e) Some physicists consider our three-dimensional universe as one with hidden extra dimensions curled up so minutely in space that we don't experience them. But we need our obvious three spatial dimensions. According to theoretical physics, the inverse square law of gravity, electromagnetic radiation, and our three-dimensional universe are all locked together mathematically. The inverse square law has therefore regulated the development of the universe. It's made things stable, including planetary orbits. The planets would not continue to orbit the Sun if the inverse square law was altered.

(f) As the late Fred Hoyle showed, if the nucleus of beryllium, the fourth-lightest element, had been more stable, stars could not have synthesized carbon, the backbone element of life. And because nature does not offer a realistic alternative to carbon, life as we know it would not have been possible. Also, elements heavier than carbon could not have been synthesized. Hoyle showed how energy conditions within massive stars ensure the high probability that carbon will be synthesized in quantity.

(g) The distribution of synthesized elements into space, where they are incorporated into generations of stars and their planets, depends on the supernovae explosions of massive stars—and on less massive stars that spew out synthesized elements. All elements above the atomic weight of iron are forged during supernovae explosions. The synthesis of iron (element number 26) marks the end for even the most massive stars because such stars cannot obtain energy from "burning" iron in the synthesis of the next-heavier element. So, when nuclear synthesis stops, a star will lose its internal supporting energy. The star then collapses on itself, and a rebound supernova explosion follows: the creation of heavier elements.

When scientists became aware of the fundamental constants, they realized that the development of the universe and life was totally dependent on them. They saw how matter is put together by the evolutionary process to produce the thousands of different molecular machines and processes that make life viable. That potential for the structural machinery of life at a molecular level existed in nature before the origin of life on Earth, and it would have existed before the origin of life anywhere, even on planets twice the age of Earth. And so the incredibly complex phenomenon of "life" develops to a stage where it can study its own origins and contemplate its own future. The potential for life exists as an innate part of the universe, and this was so from the very beginning of the universe and will continue on planets that will orbit stars as yet unborn. Life is obviously the manifestation of a multitude of processes at work within a living organism, but no one understands the origin of life or how it persists as such an amazing phenomenon. Playwright George Bernard Shaw put a handy label on it. He called it the "life force." When we try to explain the fundamental property of life, we still can't do much better than that.

LIFE'S MACHINES

During the past few decades, thanks to thousands of biochemists and molecular biologists in hundreds of laboratories, the significance of the fundamental constants can be even better appreciated. Billions of organic molecules are fit together to form the many different molecular machines that keep the simplest of living cells alive. Consider life's basic processes and start with photosynthesis, which manufactures the molecules that provide energy for almost all life on Earth. In any given plant cell, or photosynthesizing microbial cell, the molecular machinery of photosynthesis works to produce molecules from which energy can be released. And all such biochemical processes depend on the universal physics and chemistry being precisely as it is.

Another intriguing fact is that life relies mainly on the most abundant and lightest of the elements. It's intriguing because these elements (hydrogen, oxygen, nitrogen, and carbon) just happen to have the right chemical characteristics to form organic molecules with the right structures to perform life's processes. The odd element that doesn't have a role in complex molecules is helium, the second-lightest, which doesn't react chemically with other elements. Is this an aberration in the scheme of things? Not at all. The synthesis of helium from hydrogen in stars is the great energy source of the universe and the essential stepping stone to the formation of all other elements. Helium has to exist, or there would be no life in the universe. Of all elements, the carbon atom looks custom-made for life. Each carbon atom has four links to other atoms and consequently can form chains and side chains, making complex molecular structures possible. Without this structural potential, there could be no life as we know it. There is no non-carbon chemistry that could do a comparable job.

We see the ultimate complexity of this chemistry in proteins, the countless array of molecular structures formed from just twenty different amino acid molecules. Proteins are the three-dimensional tools of life and the essential building blocks in life's structure. They consist of anywhere from fifty to three thousand amino acid molecules strung together in sequences specified by the genetic machinery within cells. These chains of amino acids then form themselves into the structures needed by the living organism. That this takes place with such mind-bending accuracy and efficiency is not entirely due to natural selection, though natural selection seems to have evolved the best proteins for living processes. The same can be said about evolution and the nucleic acid molecules (DNA and RNA), which carry inheritance from generation to generation. The universal physics and chemistry make these two amazing systems possible, but the genetic code they use in every life-form on our planet came from evolution in the early days of life on Earth.

Before we leave this subject, we should consider the strangely benign properties of water. There can be no life as we know it without water. It may be an interesting exercise to try to conceive of life as we "do not know it," but attempts to do so have not been very convincing. The point for us is that we know the universal physics and chemistry supports "life as we know it" with such efficiency and robustness that it looks like a universal phenomenon, though we have no evidence that it is—or that other kinds of life could exist. So, is life as we know it just lucky that water expands as its temperature falls below four degrees Celsius, so becoming lighter and making it freeze from the top down? If water froze from the bottom up, as some liquids do, the Earth could have been uninhabitable at several points in its history; the same would apply to other planets with water.

The cell membrane provides yet another example of the universal physics and chemistry at work for the benefit of life. The cell membrane provides a protective boundary. For single-celled organisms in particular, it separates the living processes within the cell from being dissipated by the external environment. It's fortunate that the formation of membranes is inevitable because of the detergent-like nature of the molecules that form them. One end of a molecule is attracted by water but avoids oil; the other end is attracted to oil but avoids water. These molecules of a cell membrane assemble themselves automatically in water, and the best way of doing this is to form a sphere, or bubble—a cell. This way, none of the water-avoiding parts, which are inside the sphere, are exposed to water.

Thus cells of the simplest kind come into existence automatically: an inorganic start for the development of the basic unit of life. On Earth, the first proto-cells must have formed around four billion years ago, though proto-cells could have formed elsewhere in the galaxy several billion years earlier. And during the past four billion years, evolution has produced a range of high-biotech membranes, which allow the right substances to enter and leave cells. But all cells are based on the same inorganic prototype.

It also looks significant—in support of universal life—that the chemistry of this universe seems to offer only two practical polymer systems for life: proteins and nucleic acids (DNA and RNA). If we traveled to another life-supporting planet, would we find life based on those two polymer systems, but with variations? From what we currently know, it's a good guess that we would. Life everywhere would need ways to reproduce itself, so it would require an information-storage system and a way of using this stored information to produce the components for the next generation of life. This is where our genetic code comes in, mediating between the nucleic acids and the proteins, translating data held in our DNA and RNA into the proteins that life needs. The universal physics and chemistry has offered evolution a way of doing this, although evolution on different planets would be expected to produce many variations on the same theme. So, as we saw earlier, the precise genetic system on Earth should be unique.

THE GREAT MYSTERY

We now come to a great mystery. How is it that the fundamental constants, which theoretically support all forms of SETI research, are just right for the development of a universe that will produce suns and planets and give rise to life on suitable worlds? Although there is no answer yet, we had better look at the three alternative reasons for the fundamental constants being precisely as they are.

1. This universe was designed for life.
2. There exist a limitless number of universes, each with different sets of fundamental constants. Our universe just happens to have the right set for life, so we're here and aware of them. We have won the jackpot of all jackpots. Roger Penrose, Rouse Ball Professor of Mathematics at the University of Oxford, calculated the odds against the existence of our long list of fundamental constants as ten billion multiplied by ten billion thirty times. You would have to have that number of different universes (the so-called multiverse) to stand a chance of getting one with our set of fundamental constants.

Nevertheless, Martin Rees, Britain's Astronomer Royal, thinks that we should not dismiss this explanation just because it seems so incredible. He is not alone. Physicists are trying to think of ways to test the idea of the multiverse, but some feel that this explanation unloads our need for answers onto the wheel of chance. Personally, I like Einstein's dictum "God does not play dice," though he said that in response to the uncertainty principle, not to the concept of a multiverse. But the same could be said about the multiverse. Would billions of universes have to be created in order for one to have the set of fundamental constants that make life and intelligence possible? It's an uncomfortable idea.

However, the concept of the multiverse does eliminate a big problem in our thinking about a solitary universe that developed from the big bang. What existed before the big bang? Nothingness? We have a problem with this because *nothingness* means "no reality." The idea of a system consisting of a limitless number of different universes does at least provide a "hard-to-accept solution." So do individual universes come and go, while the multiverse has always existed and always will? Is that it? Is that the basic reality?

3. For unknown reasons, this universe could be no different from what it is, so that all the fundamental constants of nature are locked together in a way as yet to be discovered. This is heady stuff, and many theoretical physicists are trying to find an answer. They might succeed, although there's room for doubt for the obvious reason that all those brilliant physicists are mammals, and every species of mammal has its own neurological limitations. Some mammals are better endowed than we are for sensing the outside world, although we outstrip them all when it comes to our great capacity for abstract thought. But is this capacity great enough to understand why the fundamental constants have the precise values that make our universe possible? Maybe, but the species *Homo sapiens* can hardly be the exception among all other mammalian species

in having no neurological limitations. There must be limits to our capacity to discover and understand, but as the biologist J. Z. Young once said to me "We can only work with what is in our own heads."

THE TIMESCALE FOR SETI

The evolutionary breakthrough from normal fish to some enterprising fish living partly out of water in swampy environments 350 million years ago highlights the important biological time scale for SETI. This line of evolution led to all vertebrate animals on land, but this could not have happened if the ozone layer had not been in place. Life could not have left the protection provided by water against radiation unless a substantial ozone layer had been permanently established. *Substantial* and *permanent* are the key words here. The continuous recycling of oxygen into the atmosphere by photosynthesizing microbes and plants had to keep the whole show on the road. The fossil record of around 350 million years ago documents life's transition from water to land, and life needed a substantial ozone layer to be in place. It may have been in place earlier, but the prominent spectral line of ozone at a wavelength of 9.7 microns from the Earth's ozone layer has been astronomically detectable for the last 350 million years. And any alien astronomers near enough and looking in our direction and detecting that line (and other spectral lines that indicate the presence of life) would have known that photosynthesis on a planet out there had led to a highly evolved biosphere.

Photosynthesizing microbes and plant life (or the alien equivalent) would have to release oxygen for an eternity to form an ozone layer. It took about three billion years on Earth for organisms to produce enough oxygen, though it is believed that a great deal of oxygen arrived in comets that hit Earth. The formation of our ozone layer was slowed down because oxygen is chemically very active and oxidizes everything in sight. So most of the oxygen released by photosynthesis in the first couple of billion years of Earth history disappeared from the atmosphere and the oceans in the oxidation of minerals and rocks. Not until all these oxygen sinks had been saturated did the level in the atmosphere begin to build up.

It's estimated that by about 550 million years ago, at the start of the Cambrian period, when the evolution of many marine life-forms appear in the fossil record, the oxygen content of the atmosphere was half what it is today. But by four hundred million years ago, it was probably at about its present level of 21 percent; some later periods of prehistory have had higher levels than this. Anyway, we definitely know that a detectable ozone layer was in place by 350 million years ago because at that time four-legged animal life began to colonize dry land and they would have needed its protection to survive.

Scientists who investigate the molecular structures and functions of photosynthesis will tell you that photosynthesis is probably a universal process, though the complex molecular machinery of photosynthesis should differ from planet to planet. Actually, the chemical machinery of photosynthesis is not identical in all plants on Earth. However, photosynthesis may be a phenomenon on most life-supporting planets because it enables life to tap the major source of energy: radiation energy from the central star of a planetary system. And when photosynthesizing microbes and plants have released enough oxygen into their atmosphere, an ozone layer will form, conveniently providing a radiation shield for land life. At the same time, a spectral signal will be continuously available, enabling advanced alien life to detect the presence of life. Nature is very obliging.

ONE WAY TO LIFE?

Scientists have long considered the sea to be the most likely birthplace for life. Based on this, alien life might need planets with plenty of water to get started. But the prelife chemistry could not be spread throughout the vast volume of an ocean. It must have been concentrated in some way. The molecular bits and pieces that formed the first self-reproducing entities on Earth were unlikely to come together by chance in the primordial oceans. There had to be specific sites where the right chemistry existed for the first self-reproducing cells to form. Some scientists favor seabed hot springs, which would have existed in abundance on the seabed four billion years ago and are good candidates for the birthplace of life. They could have provided the necessary chemicals coming up from the Earth and appropriate temperatures for the chemistry of life to begin. These conditions, however, no longer exist because of oxidation: today, the level of oxygen would stop the chemistry that once led to life. There was virtually no oxygen around four billion years ago.

Today, many bacteria and archaea live in hot springs and by hot seabed vents in water up to ninety degrees Celsius in the kind of conditions that may have supported the origin of life. The theoretical attraction of such sites is that they could be common on earthlike planets, especially in their early days. There is also the conclusion, based on the fossil record, that life formed as soon as conditions on Earth permitted it to do so. The oldest fossils look like photosynthesizing bacteria, but they are in fact 3.5 billion years old. There is also evidence of substances associated with life that are 3.8 billion years old. However, not every paleontologist agrees with these findings. This is obviously a subject for SETI to get settled because if life first formed around two billion years ago, instead of about four billion years ago, we could more easily entertain the possibility that it was a chance event—that it took place after a couple of billion years of random chemistry. Obviously, if life is an inevitable phenomenon, there will be vastly more life in the universe than if it resulted from a chance event that happened on Earth after a couple of billion years.

NO DICE

When Einstein said "God does not play dice," he was objecting to quantum physics and the uncertainty principle. But the same could be said about the origin of life. Although the chance combination of organic molecules has been entertained as the possible source of life on Earth, this may seem too much like God playing dice. My guess is that when the origin of life is finally determined, it will be seen as an inevitable event, given conditions on the primordial Earth, so that it could happen on any earthlike planet anywhere in the universe. This would fit well with the theological view that the universe was made to engender life and provide homes for it, though most scientists would like to have another explanation. However, given the fundamental nature of this subject, it's odd that it has not received much attention in either theology or in SETI, the first being our oldest intellectual discipline and the second, the most recent.

BIG BRAINS AND THE ROBOTS

It's quite clear from life's history on Earth that having life on lots of planets may not necessarily lead to beings who explore other planetary systems in flying saucers or who broadcast messages across the galaxy. So, let's speculate on the paths available for the evolution of intelligent creatures who might provide some evidence for us of their existence. When we consider the billions of animal species that have come and gone—mostly gone—and the moderately intelligence species that live with us today, the evolution of broadcasters and space travelers looks far from inevitable. But it's happened here, so it's not impossible, even though it might be improbable. The basis for this assumption is that among the countless number of species that have evolved on Earth, only a few species of pre-humans have had the potential to evolve into future technologists.

If we visited West Africa five million years ago, we would find several species of pre-human apes but only one species that led to *Homo sapiens*. Why was even one evolutionary line so successful? Diet may have been a factor. All our close relatives (including the apes) are vegetarians, except for our nearest, the chimp, who, we are told by the field researchers, eats about 5 percent meat. Now way back on the African plains, there was plenty of meat, but it had to be caught. And catching fast-running meat takes intelligence. Hunting large animals is dangerous and requires planning and cooperation and communications. So a pre-human species that fancied a meaty diet experienced an evolutionary pressure to become brighter.

It's hard to envisage evolutionary pressures that would make vegetarians the rulers of their worlds. Vegetables stay put and don't fight back. You don't need planning, communications, cooperation, and weapons to be a vegetarian. We know that apes and some birds—and even the octopus—can use objects as tools, so there's a tendency to evolve this capacity. But unlike other clever creatures, our ancestors must have been able to envisage the tool or weapon needed for the task before they made it—that's the important point.

However, there are plenty of big brains around. The dolphins and whales have very large brains, bigger than ours, but with only flippers and a tail they're never going to be noted for their technology. This development of big brains in whales and dolphins shows that our big brain is not unique in the evolutionary history of our biosphere. Actually, according to the fossil evidence, the dolphins and whales had big brains before our ancestors, the upright apes who roamed the African savannas from about five million years ago. For a while, the whales and dolphins were the brainiest animals on Earth, and they evolved their brains through the need for an underwater sonar system and for social communications and cooperation. So there might be big brains on other worlds that never produced technological civilizations. For other Earths to have our counterparts in residence, a way of life must have developed that forced certain species to evolve both big brains and an ability to manipulate and exploit objects to a point where technology begins to develop.

We know roughly from the fossil record how this happened with our own ancestors, starting with the apes of about fifteen million years ago. The biological potential to rule the world was present in those apes, but there had to be factors that forced them out of their forests into a more demanding way of life. Probably the apes became too crowded as the climate changed and reduced the forested regions of Earth, and as grass took over from trees because it had evolved to survive fire and grazing. Consequently, a lot of apes found themselves on the open savannas where the demands of life drove the evolutionary development of both bigger brains and better hands for the fine manipulation of objects. That's the general view, which may or may not be completely correct, but our presence today depended on some apes being ready to evolve into technological creatures.

Coincidentally, at about the time some apes were on the evolutionary path to *Homo sapiens*, primitive bearlike creatures were returning to the sea. Those bearlike creatures evolved into big-brained whales and dolphins because of the environmental pressures of their new environment: they needed to see underwater in all conditions to find food. For this they evolved sonar systems that process sound waves so that they could see by sound as we see by light.

But this development requires the evolution of plenty of nerve cells in the networks of large brains. Visiting aliens who saw just the isolated brain of a bottlenose dolphin, which is larger and more highly convoluted than our brain, would reasonably conclude that this species was probably in charge of the planet.

We therefore have these two unrelated examples of how big brains can evolve. But are such evolutionary histories—or other, equally effective histories—going to be repeated on other earthlike worlds? Actually, the evolutionary path to whale and dolphin brains looks a lot more likely than our own. Animals might be expected to evolve sonar systems to hunt for food underwater. I suppose the same kind of thing might take place on dry land if the atmosphere was opaque enough, but then photosynthesis wouldn't be able to operate, and there would be no food to search for. Other worlds will put different evolutionary pressures on life, but for high intelligence to evolve, the right pressures would have to come at the right time, when a species was biologically ready, which was the case with our jungle-dwelling ancestors.

It's difficult to envisage other factors that could push neural development to the technological takeoff point, so I'd be interested to hear from readers on this subject. The probability of high intelligence evolving can be guessed from life's history here, but we can't assume that this is typical or that it is the only way by which a technological species can evolve. Only if the technologists on other worlds have followed a roughly similar evolutionary path to us could we expect them to look similar to ourselves.

THROWING STONES DEVELOPS THE BRAIN

If you have seen the film *2001: A Space Odyssey*, you may remember the scene at the beginning where an ape-man picks up a bone, thinks for a while, and then starts beating a nearby skeleton of a large animal. We then see flashes of the ape-man's consciousness in which he sees himself beating a live animal to death. He goes on beating the skeleton with as much dexterity as we ourselves could apply, the bones breaking and flying into the air.

But there is one thing wrong with this excellent visual depiction of mental simulation in the use of weapons. Something like eighty different muscles are needed to do what that ape-man suddenly began to do. Anatomists and primate anthropologists believe it took more than a million years for these muscles and the nerve fibers controlling them to evolve. So how did this happen? The paleontologists in this field of research have found that the first upright apes had brains no larger than those of present-day chimps, but they walked upright and so had their hands free. Some biologists have therefore suggested that powerful throwing and hitting in the quest for food, what they call "projectile predation," led to the evolution of better-adapted muscles and nervous systems. It's a fact that highly accurate and powerful throwing in sports like cricket and baseball depends on the muscular control and coordination of many muscles of the thorax and upper abdomen. The same applies to hitting, as we see in tennis and similar sports. A lot of processing power—a lot of nerve cells in the brain and elsewhere—are needed to control the muscles involved in precision throwing and accurate hitting. Obviously, better throwing and better hitting with stones and clubs and bones killed more prey and provided more dinners, and would also have provided an effective means of defense—all of which had survival value and led to better brains.

Anthropologists have found that the spinal cords of the great apes, our nearest living relatives, contain on average only about half the nervous tissue found in a human spinal cord. This could explain why chimps and gorillas are not good throwers or hitters. *Homo erectus* might have been somewhat better, though probably not up to our sporting standards. Anthropologists have also suggested a link between projectile predation and the origin of human language. The evolution of nerve cells in the spinal cord and thorax needed for good throwing and hitting also made possible the fine muscular control essential for human language. Our ancestors might have thrown and hit their way to articulate speech. Anyway, it's a good bet that predatory throwing and hitting were common ways to get a dinner—large sticks and stones being freely available and potentially lethal. And they would presumably be available on other Earths for any creatures that were able to use them.

Geneticist Steve Jones once said, "The beauty of paleontology is that no one can disprove your best guesses." And the link between the evolution of big human brains and the origin of language through precision throwing and hitting is a guess, but it's a good guess, and it's based on anatomical studies. Of course, the best brains would also have coped better with emergencies in a dangerous environment. Way before the time of *Homo sapiens*, young humans and pre-humans would have met life-and-death situations long before they reached reproductive age, and those with the best brains survived such emergencies to pass on their genes. In this way, the human line evolved spare neurological capacity. And it's that spare capacity that we now use to try to understand life and the universe.

EXOTIC LIFE

Before we leave this chapter, we should consider exotic life: *life as we do not know it*. Some scientists have speculated freely on exotic life, and some science fiction writers would be penniless without it. For all we know, the universe may be bursting with exotic life, though chemistry is against this possibility. Silicon life has been the brand leader in exotic life for many years because the element silicon is in the same group as carbon and can form chains of molecules. But the silicon bond is attacked by both water and ammonia, so silicon life that landed unprotected on Earth could melt away. Another major objection is that silicon would be no match for carbon in the extremely complex molecular structures and metabolic processes that make life as we know it possible. It therefore seems unlikely that life would form using silicon instead of carbon.

There's been plenty of speculation on life even more exotic than silicon life: hot plasma life, neutron life, frozen hydrogen life, and pure energy life, to name but a few that have originated in the heads of scientists and science fiction writers. So although the nature of extraterrestrial life is open to speculation, we can hardly deny that "life as we know it" has been so successful on Earth that it could be a universal phenomenon. What's good enough for Earth looks good enough for the whole universe where conditions are right. Life basically similar to ours may therefore exist out there, although no one can deny the possibility of exotic life, which leaves science fiction writers with plenty of room to create aliens that are both fantastic and almost credible.

However, the problem with sci-fi life-forms, especially those on television, is that although they may look all right on the outside, they remain a big mystery on the inside. Alien creatures come fully formed into the sci-fi universe, and no one knows how they evolved or how they function. Take the ultimate in unbelievable fictitious carnivores: the aggressive creatures in the *Alien* films. How did they evolve dentition good enough to eat a *Tyrannosaurus rex* for breakfast? Those terrifying teeth didn't evolve from a diet of astronauts, but there didn't seem to be anything else in sight for big hungry monsters to eat—and large carnivores need plenty of prey in order to evolve. All aliens, wherever they might be, will have been shaped by the lives of their ancestors and the environments in which those ancestors had to survive. Animals don't have terrifying teeth unless their ancestors needed terrifying teeth.

It's the same with brains. We have big brains because our ancestors needed big brains to survive. This applies to all life everywhere in the universe. All technological creatures, our counterparts, will have a biological history in which big brains played an essential part in their survival. And since all biological histories are going to be different, the evolutionary products from those histories are also going to be different, and this includes brains. Alien brains may be similar to ours, but they will be different. Whether or not that difference would make communication impossible, we cannot know. And we may never know. The aim of discovering the signatures of life is not to open congenial communications with alien communities but only to confirm evidence that alien life and intelligence exists. That's what we may do, and it's all we can reasonably expect to do.

CHAPTER 5

WHERE ARE THEY?

Many people argue that if UFO reports were true, the aliens would have introduced themselves by now. Unless we believe the stories of contactees and abductees, this has not yet happened. The aliens have made a wise decision. Apart from the obvious chaos they'd cause, visiting a strange biosphere wouldn't be healthy for aliens if they were biological beings. It wouldn't be like our earthly explorers arriving on unknown continents of our planet. Those explorers didn't know what monsters might dwell in lands across the seas. As it turned out, microbes, not monsters, caused all the problems, although the explorers didn't know about microbes at the time. They discovered plenty of humanoids of the species *Homo sapiens*—another fact that not everyone appreciated at the time—and they were able to communicate, shake hands or whatever, eat the local food, and make themselves at home in strange lands. For reasons we've already considered, no biological aliens could explore and exploit other inhabited planets in that way. Robots could do so, of course, but since they would have come from advanced civilizations, they might bring with them some sort of "noninterference directive," especially because planetary biospheres like ours are very rare, and highly intelligent species and their civilizations are even rarer. That might explain why there is no evidence of alien interference on Earth: we're a very rare lot.

All life on Earth needs to be conserved. Although the aliens have never attempted to colonize our planet, this hasn't deterred some scientists from writing articles about future expeditions to establish human colonies on other inhabitable worlds. And *Star Trek* crews have gone through several television series regularly visiting other worlds and socializing with the inhabitants. It may be good drama, but it's impossible biology.

The vastly different time frames in which evolution must take place on different worlds is a formidable obstacle to meeting our equals in space and time. A million years is a tick of the clock in the history of planetary biospheres, and world civilizations will not emerge at roughly the same "tick," so there will be no "equals" in our galaxy, a point that all *Homo sapiens* who hope to talk with aliens should understand. This is where the well-established principle of mediocrity, which we reviewed earlier, has led SETI scientists astray in their search for extraterrestrial messages. We live on an average habitable planet warmed by a common type of star, and four billion years is probably enough time to evolve humans to the level of SETI astronomers. But technologically savvy creatures on other planets might experience the same increasing rate of advancement in science and technology that we are. Our knowledge and application of science advances faster and faster, driving technology forward and changing the world, while we are constantly called on to master new technologies. The present rate of development can't continue indefinitely, yet we can expect that civilizations far older than ours would have science and technologies far more advanced and therefore incomprehensible to us. If we accept this, then an extraterrestrial presence could likely display phenomena that we can't explain. Therefore, since we can only guess where aliens are or what they might be doing, we should be on the lookout for any strange and unexpected phenomena.

Most SETI scientists have not accepted this position. They have tried to guess what aliens might do on the basis of what we can do—at the present time. Anyone should be able to see that this approach is not sound; it's the kind of thinking that has led us astray in the past. So, to remind ourselves of the basic situation, let's go into sci-fi mode for a moment and consider the numbers astronomers provide for the ages of the oldest stars. Since at least some of these stars probably have planets similar to Earth, it's possible that the first civilizations in our galaxy developed around four billion years ago. If they did, our civilization could be part of a galactic community unknown to us. This is sci-fi speculation, of course, but judging from what we know, it's not impossible.

TEA WITH THE ALIENS

At least one thing seems certain. If the aliens have made it to the solar system, they're not going to drop in for tea and cakes any time soon. However, they have been invited—maybe not for tea and cakes but to communicate with us in some way. One such invitation, supported by eighty scientists and academics, has been put on the web so that the aliens can send us an e-mail. This "invitation" may seem hopelessly unrealistic, but it is not as unrealistic as the lawyers who are so wedded to the principle of mediocrity that they have drawn up treaties in preparation of discussing diplomacy with the aliens. These lawyers don't worry about how they will negotiate with creatures ten times brighter than they are; never mind the ultimate robots that might be even brighter.

Nevertheless, although normal human-to-alien contact seems impossible, we can speculate that some of the aliens' motivations would be the same as ours, otherwise they would not have developed civilizations and space technology. The desire to know more necessitates exploration, and no species is going to reach the technological heights without a powerful drive in that direction. The motivation for aliens visiting the solar system would be discovery. They wouldn't come here for a change of diet or to upgrade their DNA, even if their own reproduction was controlled by such a molecule.

Colonization is a natural development on Earth because as we multiply, we have to spread out. But biological beings can't be expected to spread into other biospheres. Apart from troublesome microbes, hungry aliens would have no food to eat. Differences at a molecular level would make our food useless or highly toxic. We may see *Star Trek* "away teams" dining with aliens and eating the native food, but no real biological astronauts will enjoy such hospitality.

This lack of visitors leads to the view that there aren't any intelligent aliens. Maybe dumb animals, but not anyone bright enough to build interstellar spaceships. The reason? They've had so much time to reach Earth, they should have been here long ago. Therefore, if they're not in town shopping at the supermarkets, they don't exist. Of course, several things could have delayed the delights of shopping on Earth, such as being hit by a large meteorite, global warming or cooling that got out of control, nearby supernovae, or gamma rays from nearby astronomical catastrophes. But during the four billion years available to visitors to the solar system, not everyone is going to be extinguished by a global or astronomical catastrophe.

However, those who have argued that extraterrestrials do not exist because they are not here have neglected the reasons for this. As we've seen, differences at a molecular level in the plants and animals we eat might do more than upset alien stomachs, if they have stomachs. The organic molecules in our food wouldn't be the same set as the aliens are accustomed to. The organic molecules in the food we eat have evolved, like all other parts of life. And no two biospheres will have the same evolutionary history. So unless the aliens could synthesize all the organic molecules they needed, especially the amino acid molecules to build their proteins, they'd be in trouble. In a real-life situation, humans who might visit other Earths would be in trouble because humans can synthesize only about half the amino acids they need. The other half has to come directly from earthly food, from things that were once alive. Of course, none of this would apply to nonbiological beings (robots) who might run entirely on batteries that could be recharged anywhere in the universe.

THE BROADCASTING GIVEAWAY

When queried by people who fear that SETI activities might reveal our presence to unpredictable and dangerous aliens, the SETI scientists reply, "We've given away our presence already, so SETI won't change a thing. Astronomers on neighboring worlds could see our television transmissions as bright as the Sun at certain frequencies." Certainly, an expanding sphere of radio and television broadcasts moves outward at the speed of light, but these signals are such a recent phenomenon that they're completely insignificant as far as attracting alien attention is concerned. Alien astronomers could have detected the spectral lines from our atmosphere, the fingerprints of life, at any time during the past 350 million years, and probably much earlier. These lines have certainly been present since animals colonized dry land; that is clear from the fossil record. Yet the connection between our biosphere and its telltale atmospheric spectral lines has not figured in speculations about the possible activities of intelligent aliens. This is surprising, because the detection of planetary spectral lines has great significance for SETI research, not because we might find aliens in this way but because they might find us.

Before launching their probes, alien visitors would have found the most interesting planetary systems in their neighborhood—something our astronomers are doing today. But the main technique our astronomers are using depends on the effects of gravity, so the planets discovered so far have mostly been larger than Jupiter and orbit near their alien suns. The nearer a planet is to its sun and the more massive it is, the greater the gravitational effect on that sun, although the orbital periods calculated for large planets range from around three days to thirteen years.

The discovery of massive planets with very short orbital periods was a great surprise. This is not what astronomers had expected. Theories of planetary formation led us to believe that the solar system would be about average in its configuration—the old principle of mediocrity at work again. But the current picture of planetary systems being dominated by massive worlds could be misleading. Earthlike planets may exist in abundance because there are so many planetary systems in our immediate region of the galaxy. What astronomers hope to find are planetary systems similar in form to the solar system where large planets, like Jupiter, orbit far out from the central star (sun) and protect small inner planets from incoming space debris. The great gravitational pull of Jupiter has been protecting Earth in this way since the dawn of life—although, based on the state of the Moon, Earth must have taken a hammering from meteorites in its early days.

THE OZONE GIVEAWAY

Our ozone layer formed long ago during the early days of the Earth, but it likely reached its present level several hundred million years ago. Its formation depended on the presence of oxygen in the atmosphere: the more oxygen, the more substantial the ozone layer. And it was photosynthesizing microbes and plants that released that oxygen during their metabolic processes, creating the sort of atmosphere we have today. But oxygen is a very reactive gas, and for at least a couple of billion years, it was lost in the oxidation of the Earth's rocks. Some scientists have suggested that comets hitting the Earth supplied an additional source of oxygen, and this is probably true. But the important point is that once a substantial ozone layer had formed, it provided protection against lethal radiation.

The evolution of life on dry land and the development of a complex biosphere depended on this protection. If astronomers find the prominent ozone line (at around 9.7 microns) among the spectral lines of another planet, they would probably have found a biosphere in an advanced stage of evolution, one where life had been able to leave the protection of water for a more advanced existence on dry land. The ozone line would be very significant, although other spectral lines from planetary atmospheres could provide information. This type of astronomical research, as more advanced telescopes are produced, should eventually determine the abundance or lack of life in our galactic neighborhood. And this is both a possible and probable development for any planetary civilization that is trying to discover its status in the universe.

FINDING MORE DISTANT WORLDS

In January 2006, leading science journal *Nature* published a remarkable research paper to which seventy-three scientists in twelve countries contributed. That number of contributors to one research project suggested something important had occurred—and indeed it had: the discovery of the first earthlike planet beyond the solar system. This planet, however, was unlike others already detected in neighboring planetary systems ("neighboring" here means within a few tens of light-years.) As we've just seen, these massive planets— like Jupiter and much larger—were discovered first because the detection method used depends on gravitational effects. A massive planet affects the movements of its sun in space and so shifts the frequency of light coming from that sun. Astronomers record these changes and from them calculate details of the planet. Some massive planets orbit close to their suns, which greatly increases the detectable gravitational effects, but planets like Earth will not show detectable gravitational effects because they're not massive enough. Such planets can't be discovered by this gravitational calculation, so astronomers developed a new technique to do what seemed impossible: they discovered the first earthlike planet outside the solar system, a planet far more remote than any planet previously detected. It's a rocky planet—like Earth, Mercury, Venus, and Mars—but it is twenty thousand light-years away. And it's horribly cold because the star it orbits is much cooler than our Sun.

So how did astronomers find such a distant world? The answer: by gravitational lensing. They took advantage of the fact that a strong gravitational field bends and focuses light rays like an optical lens. In this case, a background star had come into alignment with a foreground star. The foreground star is the one with the planet in orbit, and such an alignment makes gravitational lensing possible. Light rays from the background star naturally spread outward, but as light passes the foreground star, its gravity bends the rays inward like a lens. The gravity of the orbiting planet also bends the light rays just a little. And it's this added effect provided by the planet that the astronomers detected and from which they were able to calculate the mass and orbit of the planet.

The size and brightness of the star plus the position of the planet's orbit make this world far too cold for life as we know it; nevertheless, the media made this distant world a front-page story. One newspaper story said the discovery was a quest to find evidence of life and intelligence beyond the Earth. In a way, this is correct, although no direct evidence of life could be detected at such a distance, even if the planet was warm enough. However, if astronomers can detect many small, rocky planets in this way, they could conclude that earthlike planets are abundant, making it statistically probable that some would possess the kind of conditions that support life on Earth. And the fact that the first planet detected by gravitational lensing is a rocky planet, rather than a massive gas giant, may indicate that rocky planets are as common as the planetary giants.

Searches to exploit gravitational lensing have so far been directed toward the center of our galaxy where stars are most abundant and most likely to come into alignment. Such alignments are essential. However, these inner galactic regions are not exactly ideal locations for a few billion years of undisturbed organic evolution. It is much better to live way out in one of the spiral arms—like we do—far from exploding stars and black holes. This astronomical research is a great contribution to SETI because it should lead to an estimate of the potential real estate available for the evolution of life-forms capable of developing technology and civilizations. Some SETI astronomers will one day be targeting results from gravitational lensing to search for signals from other worlds.

Recently, I was looking at the Hubble Deep Field, an image that gives an idea of the amount of real estate available for life out there. To create the image, the Hubble Space Telescope made a series of observations of a section of the sky just 1/140 that of the full Moon. In this tiny area are about three thousand galaxies of stars. This is amazing when we consider that our galaxy alone contains more than a hundred billion stars. So we shouldn't be surprised if someone somewhere out there is bright enough to find a way of conveniently crossing the light-years.

THAT WIDE WINDOW
OF OPPORTUNITY

Unless interstellar transport could be developed from physics as yet unknown to us, the solar system may never have had visitors, even from the nearest planetary systems. With a window of opportunity a few billion years wide, though, it's hard to believe that the Earth has remained undiscovered. Already we have seen how earlier civilizations could have found us, and we've seen how our own astronomers search for biomarkers from other planetary systems with a good chance of detecting life in our very limited region of the galaxy. The European Space Agency's Darwin space telescope and telescopes being developed in the United States will study the spectral signatures of other Earths within about thirty-five light-years. Telescopes will be built in space and transported to the outer regions of the solar system, where observations will be least affected by interplanetary dust. What can be detected will at first be limited to the nearest planetary systems, but the technology and the science involved is bound to advance. Of course, the relevance of this for our subject is that it shows what science and technology allow us to do in our time. And what we can do in our time, other civilizations may have achieved millions or billions of years ago.

So it seems that astronomers rather than biologists will be testing the major question in biology. If life exists in the very small region of space that Darwin and other telescopes can observe, it would follow that life can exist in similar regions of our galaxy of more than a hundred billion stars—and in the billions of other galaxies astronomers are studying. Yet finding evidence of life is not the same as finding evidence of intelligent life and civilizations, which is what SETI astronomers have been searching for during the past fifty years. Of course, when planet searchers find planets with biospheres, SETI astronomers will start scanning them for intelligent signals, especially if planets are found with suspicious spectral lines that might come from technological activities. But even if an alien civilization has active technology, it doesn't follow that it's using radio to transmit signals that we could detect. Alien civilizations may have come and gone during the past four billion years, so, statistically speaking, there would have had to have been a continuous series of radio transmitters in action during that time for detectable radio signals to be present in our time. So, SETI astronomers will have to be exceedingly lucky to succeed; although, in their favor, there are a few billion sunlike stars in our galaxy alone, and, as they say, there's strength in numbers.

This biosphere-hunting research is well based in science and seems sure to provide new information. But confirming the existence of intelligence and civilizations in this way may not be possible, and even if it were, it would take a long time and be very expensive. Our only chance of making a low-cost breakthrough in our time depends on finding evidence in the solar system, either archaeological evidence, perhaps millions of years old, on the Moon or Mars; or present-day evidence that something in the UFO phenomena could confirm the presence of alien artifacts.

The probability that such evidence might exist obviously depends on how easy it is for advanced civilizations to cross interstellar space, though we can see from our own advances in artificial intelligence that advanced robotic technology could make interstellar spaceflight a lot easier. We also have to remember that some of today's distant stars were in our neighborhood in the past. All stars and their planetary systems are on the move, continuously circling the galactic center. Many that are now far away have in the past been relatively near. Thus the civilizations near enough to detect a life-supporting planet in the solar system and near enough to travel here, or to send intelligent probes, would not have been constant throughout galactic history.

From all this information, we can create a sci-fi scenario that might be a reality. Imagine you're a superintelligent ET with the ultimate in astronomical and space technology, and you're exploring your neighborhood of the galaxy. There are a thousand sunlike stars within a hundred light-years, so, having a long lifespan and a lot of spare time, you want to launch robotically controlled spacecraft to explore the most interesting planets. Your main interest, as ours would be, is to study the nearest planets with life—"blue planets," like the Earth—and to see if life on any of these planets has created an active civilization. Your astronomers have already identified those planetary systems with life; the rest appear to be sterile because their spectra lack any biomarker lines. But those planetary systems showing biomarkers tell you that if your spacecraft reach their targets, there will be planetary biospheres to explore—maybe not civilizations, but plenty of interesting life.

Smart aliens will know all about this and will proceed as we would. They would not have to launch a thousand probes to find one planet with advanced life. They could just go to that one planetary system and leave the other 99.9 percent, knowing that an advanced biosphere awaited them. That this opportunity exists is the best support for the hypothesis that some evidence of alien artifacts could be in the solar system. It doesn't mean that the aliens are here in our time or that they have ever been here; it only means that we can see how they could have found Earth and visited us if they were able to cross interstellar space. So we're making some progress in our thinking about alien visitors—probably probes or whatever the ultimate in this field of technology might be.

In a NASA report in 1977, John H. Wolfe, who worked on NASA's SETI program at the Ames Research Center, voiced the official attitude about probes at the time: "To 'bug' all the sun-like stars within 1,000 light-years would require about 10^6 probes. If we launched one a day this would take about 3,000 years and an overall expenditure well over $10 trillion. Interstellar probes are appealing as long as someone else sends them, but not when we face the task ourselves."[1] In other words, no one, not even super-civilizations, is going to go to such trouble and expense. It's the same old story. Unconquerable problems look easy after one finds the answers. We already know how aliens could have found the solar system among thousands of planetary systems without life, though we do not yet know how they could have gotten here. But this is probably a mystery only because we don't yet know enough physics—just like NASA's need for a million probes in its 1977 report.

TIMELESS TARGETS

What SETI needs in order to discover aliens are timeless targets rather than transmitted signals from distant worlds that might last briefly in the vast time frame in which such signals are theoretically possible. As long ago as 1959, physicist Frank Dyson suggested that we might be able to detect the presence of advanced civilizations from their engineering activities carried out on an astronomical scale. Dyson thought that civilizations like ours would pursue two main objectives: they would want to maximize their living space and maximize their use of available energy. On Earth, we experience a similar problem: we live on one planet in a planetary system where almost all the Sun's energy is lost into space. The Earth traps only one-billionth of the energy radiated by the Sun. We can't do much about that at present, but if some advanced civilizations also live energy-hungry lives, they may have done something to capture more energy from their suns. Dyson's suggestion devised a way to trap most of a star's energy while also greatly increasing a civilization's living space.

Civilizations could have more living space and more energy by using materials from surplus planets and moons, plus a lot of asteroids, to construct a vast shell of separate habitats that would orbit the central star and soak up most of its radiant energy. The idea was that a continuous shell of only several meters could be built that would totally enclose the central star. It's hard to imagine the super brains of the galaxy having the mentality of worker ants and completely reconstructing their planetary systems in this way, but this is how Dyson spheres were portrayed in the media and in science fiction. Calculations soon showed, however, that a solid shell was not an option. Stresses would fragment the immense sphere as it revolved to produce gravity for the internal inhabitants. Pieces would then drift into the central star to produce a very unhappy end to an over-ambitious civilization.

Actually, the builders would never get as far as a complete sphere if they used lifeless planets like Mars, Venus, and Mercury for building materials. They would thus disrupt the gravitational balance of their planetary system and send some of its worlds disappearing into outer space. Nevertheless, a sort of sphere formed by many separate space habitats constructed over a long period does look feasible, providing the builders had a mastery of spaceflight to transport materials and colonists safely to various regions of a planetary system. There would also be the problem of keeping all the space habitats livable, which would mean maintaining self-sustaining ecosystems if the inhabitants were biological.

If our own world civilization developed in this way—and it may have to— *Homo sapiens* would need very stable biological ecosystems to survive in space colonies. The biosphere of Earth has a vast number of interlocking ecosystems involving millions of different life-forms, so things can become unstable in some parts and later recover or change without upsetting the whole. This does not apply to small habitats, as the failure of the two sealed Biosphere projects in the United States have shown. When the ecosystems of each self-sustaining structure failed, the human inhabitants had to give up and walk out the door, which might be rather dangerous if you were in orbit at the time. With many orbiting space colonies in the future, the failure of their ecosystems might be a big problem. The call might go out: "Our ecosystem has broken down. Can we live in your colony until it's fixed?" However, if *Homo sapiens* ever build a Dyson sphere of space habitats, billions of people could have all the energy they needed. They could run all their technologies and pollute freely for the foreseeable future—providing they didn't mind living in space. And many might prefer a rich life in space to a poor one on Earth.

The most advanced species in the galaxy would not wake up one morning and decide to put the robots to work to demolish all their old planets and build an enormous sphere around their sun. There is a more natural way by which an orbiting sphere of space habitats could come about. We see this from the work of the late Gerard O'Neill, once a professor of physics at Princeton University. With his collaborators in the United States he showed the feasibility of mining the Moon and asteroids to build space habitat—although a safe, reliable, and cheap space transport system would be needed, and the building would be done over centuries, not as a single project. It's not unrealistic to envisage a faraway future in which a sphere of some sorts has developed over time through the building of space colonies that would eventually become so numerous they would absorb a large proportion of the Sun's radiation. And if, as O'Neill showed, such space colonies are a possibility for our future, we could hardly expect to be the first to build them. This seems such a necessary development for energy-hungry civilizations that once they had space colonies in their own planetary systems, they might use their skills in other planetary systems to provide whatever types of habitat were needed.

Gerard O'Neill's calculations demonstrate that the building of space colonies could be an inevitable development. He showed that beyond a certain point, the Earth's surface is an unsuitable place for technological activity. The stresses on our planet would be too great. The options would be to stop technological growth or to go into space where free energy is unlimited, where we could expand, manufacture, and consume in an unlimited way, and where global warming would be no problem. O'Neill calculated that the solar system offers the resources to allow our civilization to continue to grow at its present rate for the next five thousand years. This might not be what everyone would want for the future, but the growth of civilization and technology looks unstoppable.

But the point for our discussion is not what future *Homo sapiens* may do but how to make a credible guess about the Dyson-O'Neill spheres that could have developed slowly in many planetary systems during the past four billion years. Can we find such spheres that our counterparts built long ago, even though they are now only relics of long-extinct civilizations? We'll see later.

What we have here is a difficult hypothesis, one needing far more advanced astronomical technology to really test it. Astronomers in the near future looking for evidence of Dyson-O'Neill spheres should find them as close to the solar system as possible. The few astronomers who have tried to find evidence of these spheres have had a problem. The infrared radiation that Dyson's hypothesis predicted was similar to that received from the discs of dust and gas orbiting some stars. It was not possible to differentiate this from radiation absorbed by a sphere of habitats and re-radiated in infrared frequencies. So even if the greatest minds in the galaxy ended up living in Dyson-O'Neill spheres, we might not be able to tell their creations from discs of particles and gas. That was the initial reaction, but it didn't go far enough.

WHAT TO LOOK FOR?

Only one person, as far as I know, has found possible Dyson-O'Neill spheres: Charlie Conroy of the University of California, Berkeley.[2] He first used a catalog of a thousand stars compiled by Geoff Marcy and Jason Wright, one of the world's most successful planet-hunting teams. Their catalog listed mature stars in our neighborhood that might have planets in orbit. They were not interested in young stars with orbiting discs of matter—the discs from which planets form. And, as we've seen, infrared radiation from such discs could mimic Dyson-O'Neill spheres.

The next stage in Conroy's research was to consult the two major catalogs of stars found by satellite telescopes to be radiating significantly in the infrared. Here Conroy found 539 stars that also appear in Marcy and Wright's catalog. The next step was to work out what would be a natural level of infrared radiation from such stars, and he found that thirty-three stars did not fit his graph. They showed a surplus of infrared radiation in the form of a hump toward the end of the graph. This interesting discovery induced some SETI astronomers to consider observing these stars for intelligent signals, but here again, we can see the "missing technology" problem. Civilizations capable of building and maintaining vast spheres of habitats and manufacturing units around their parent stars should have systems of communication incompatible with ours. However, the frequencies radiated by the atoms and molecules produced by technological activities with such spheres would be the same for everyone for all time, so future astronomers with powerful telescopes sited on, say, the other side of the Moon might find frequency lines in the electromagnetic spectrum that would not normally be present and that might indicate activities within Dyson-O'Neill spheres.

An obvious point to come from this subject is that civilizations in the habit of building spheres of space colonies would have rather good space transport—good enough to visit other planetary systems, at least in craft controlled by robots. Consequently, if we have any such spheres in our neighborhood, or if any have been near Earth in the past, as the stars travel in their orbits around the galaxy, there might be evidence on our own doorstep. And it could be a lot easier testing this hypothesis than checking on the reality of those hypothetical spheres, even though they may look like an inevitable development for expanding technological civilizations.

THE GUESSING GAME

We're guessing what the aliens might have done based on what we have done and on what we think we might do in the future. And much of this depends on the big unknown: does the universal physics and chemistry offers a way to practical space transport? At present, the technology of rocket propulsion is suitable only for sending satellites into orbit and probes to explore the solar system. Using it for crewed trips would be dangerous and enormously expensive. Routine crewed flights to the Moon and Mars, which would initiate entry into a real space age, can only follow a revolution in propulsion technology. Technical journals have looked in theory at nuclear fission, nuclear fusion, propulsion from the Sun's radiation (sailing to the stars on the solar wind), even matter-antimatter annihilation, a technology that takes *Star Trek* around the galaxy. In theory, antimatter would provide the ultimate in energy efficiency because matter-antimatter annihilation turns all matter into energy, while the nuclear fusion of the hydrogen bomb transforms only 0.7 percent of matter into energy.

Apart from propulsion by the Sun's radiation pressure, these methods of interstellar transport are too dangerous and impractical. Something new is needed. And that "something new" can be found only in new advances in physics. We need something like the revolution in nineteenth-century physics, when Michael Faraday and other scientists were discovering the fundamental aspects of electricity and magnetism. At the time, scientists didn't know where their work might lead. When Faraday was asked what use his research on electricity and magnetism would be, he answered by asking "What use is a baby?" Faraday's "baby" grew up to embrace the world with electronics. Today, the most expensive research explores the nature of matter and the fundamental forces of the universe with giant accelerators costing many millions of dollars, colliding nuclear particles at near the speed of light. So at delivery, the "baby" had better be a large one, considering the amount of nourishment this research has received and the length of the gestation period. And it would be a large one if it shows how matter acquires the quality of mass, which is a major subject in present-day physics.

This might lead to a propulsion system that would make a real space age possible, since mass is directly proportional to inertia and we are constantly using energy to overcome inertia—from making cars move on land to launching rockets into space. So it's not surprising that the major questions in physics today are "What is mass?" and "How does gravity interact with mass?" Is there a mechanism deep in the heart of matter that creates the ubiquitous quality of mass? Because the inertia of an object is proportional to its mass, we can imagine the benefit of, say, reducing the mass of a spacecraft by 99 percent, making it 99 percent easier to launch into space. Naturally, physicists are keen to fully understand the nature of mass and gravity, and, like Faraday, they don't worry much about the future applications of their work. At present they can tell us only about the so-called laws of nature, which describe the behavior of mass and gravity. But when they understand the mechanisms that underlie the phenomena, it may be—just may be—possible to take control of mass and gravity, making spaceflight affordable and routine.

A real space age can come about only from major breakthroughs in fundamental physics that enable us to explore and exploit the whole solar system as well as transforming transport and energy-generating systems on Earth. Colonies could then be built on the Moon and Mars, and we could have large space habitats in Earth and solar orbits—maybe even a Dyson-O'Neill sphere of our own. Spaceflight could become as easy as airline flights today, as we travel almost massless through the solar system. Will this ever be possible: massless astronauts in massless spaceships eating massless food and drinking massless drinks? How would our bodies and metabolisms cope? Not well, I suspect. And the whole technology would be vastly more complicated—and probably different—than we can imagine.

All this speculation offers a way to explore possible explanations for the reported flight characteristics of flying saucers (strange aerial craft) that are generally consistent with aerospace engineering and future potential physics. If these craft are real (and there's no proof that they are), they defy gravity and show us what is possible. Thousands of reports describe the flight characteristics that could be displayed if the effects of mass and inertia were under control. In flight, at landing, and at takeoff, the saucers seem to be unaffected by the Earth's gravitational field, at least according to the reports. Saucers have often been reported as skipping through the atmosphere, rather like a stone thrown across water that skips or bounces on the surface. Could it be that saucers in flight have such a low mass that they are bouncing on layers of the atmosphere? The "falling leaf motion" is also consistently reported when witnesses describe saucers that are hovering or descending to land. This would indicate that saucers have some mass, like a falling leaf in the atmosphere. Presumably mass is steadily increased because saucers are said to land with a bump, leaving impressions in the ground that appear to be made by the weight of many tons. Yet when saucers depart, they often accelerate out of sight in seconds. Such departures, if real, should be accompanied by a big sonic boom. That none have been reported would mean that the saucers have some way of noiselessly pushing the atmosphere out of the way. With our technology, not only would there be a big bang but the energy needed for such acceleration would be so great that no witnesses would exist afterward to describe the event. The reported acceleration would also produce enough g-forces to crush the craft's occupants to pulp if they were biological entities. It's interesting that the UFO phenomena just might be an indication to us that nature allows gravity-control technology, and that all we need do to develop it is to move up a gear or two in our grasp of fundamental physics.

Actually, the seeming immunity of alien craft to the creation of g-forces fits snugly into Einstein's principle of equivalence, which is contained in the general theory of relativity. Einstein showed that the quality of inertia can be given to matter by either a gravitational field or by acceleration. What is gained through these two means (gravity and acceleration) is apparently the same. In a higher gravitational field than the Earth's, objects would have greater inertia and more mass. More force would be needed to move them. In a spacecraft accelerating at takeoff, the mass of an astronaut increases, and it becomes more difficult for him to move because his inertia increases with his increase in mass. TV shows and films have shown this effect of acceleration on pilots and astronauts.

Witnesses commonly report that saucers make acute turns at great speed, maneuvers producing g-forces that would rip apart our level of spacecraft technology. However, a right-angle turn at any speed might be possible if the craft and its occupants were without mass and therefore free from inertia, perhaps in their own protective inertial frame and isolated from outside forces. Something like this has to be in operation if witnesses' reports are true. Alternatively, are the witnesses all reporting the same details because they've all been reading and hearing about the same stories in the media? We have to allow for this possibility. But as we'll see in the next chapter, the details of those reports have provided some data that make scientific sense, although most of the witnesses would not have been aware of this.

Flying saucer manufacturers may or may not exist, but if they do, they have based their work on physics yet unknown to us. No other branch of science could provide what is needed; that much we can be sure about. Of course, physicists may one day have to conclude that they've reached the bottom of the barrel as far as finding exploitable knowledge is concerned, that no present or future Faradays will see their "babies" grow into technological giants. If such a shortage of applicable science pertains to us, it would pertain to world civilizations everywhere. There could then be no visiting aliens, and *Homo sapiens* would never go forth to discover those "brave new worlds." However, if it is possible to use the phenomena of gravity and inertia for the purpose of spaceflight, it could be a very convenient answer to the problem of propulsion, the force of gravity being freely available wherever you happen to be in the universe. You wouldn't have to carry it with you—or pay for it. And you wouldn't need the wealth of a nation to go into space.

There is new thinking in physics that in theory manipulates time and space and that could provide even more exotic space-travel scenarios. Should this physics ha a reality to be discovered, its application might lead to the magical space travel depicted in sci-fi movies. It all depends on the height of "ceilings" in this universe, on the amount of applicable knowledge that nature can provide. I suppose it also depends on the capacity of the human brain to discover and use that knowledge, although improving the capacity of the human brain to deal with the universe is perhaps something we can leave to the future. Nevertheless, on the basis of science we currently understand, we can explore our present position—or predicament.

CEILINGS

The idea of "ceilings," which I developed in a SETI book in 1989, is fundamental because ceilings to knowledge govern where our counterparts might be and what they might be doing. The unanswerable question this concept poses is, What level of ceiling exists for knowledge that can be applied to produce new technologies? Clearly, the kind of technologies that might be possible, including interstellar spaceflight, will depend on the answer. So what exists in this universe: a high ceiling or a low ceiling or something in between? The predictions from some physicists might lead the general public to believe that our imminent comprehension of the whole universe is at hand. But for this to be possible, the ceiling to knowledge in the universe would have to be rather low—not much more than the human brain can comprehend at present. One only has to read some of the books by leading physicists to get the impression that they think their brilliant brains are capable of understanding everything there is to know about the universe—and, moreover, that they are on the verge of doing so. This may help to sell their books, but it also displays an optimism that can be dispelled by a little observation of our fellow mammals.

We do not think it strange that the chimpanzee cannot comprehend the universe, so why should we expect its more intelligent cousin to do so? In all species of mammals, we observe distinct limitations in awareness and under-standing because of the ways in which their nervous systems have evolved. Therefore, to maintain that *Homo sapiens* is the one and only mammal that is an exception among all other species of mammals is to hold fast to a pre-Darwinian view of the human brain. I'm sure they don't mean to be so antiquated, but some physical scientists do seem to think that the capacity of their brains is unlimited. One leading SETI scientist has actually written: "I don't believe the human brain is limited in any fundamental way. I think it can emulate the power of any intelligence we may find in the universe. And I expect the discovery of extraterrestrial life to bear me out on this account soon." Well, good luck—but the best brains in the galaxy might have a different opinion.

For all we know, it may take a brain many orders of magnitude greater than ours to comprehend the fundamental nature of the universe. Can we imagine that our level of human intelligence, with all its obvious weaknesses, will be working away to try to understand everything in a hundred thousand years' time? We could reasonably expect our neighborly aliens to be at least that far ahead of us in time—and maybe in science and technology as well. Our descendants in a hundred thousand years (if any there be) will either be back in the caves or vastly superior biologically and intellectually and capable of understanding the universe in ways we could not begin to comprehend. We therefore have a situation where many nonscientists believe that science is about to explain the universe, when the truth must be that we'll never have anything like a final understanding. The theory to explain everything is a physicist's dream. Only a human understanding is possible. It looks as if we would have to be made exactly in God's image with a central nervous system to match for our hopes of a complete understanding to be realistic.

The only way out of this position would be the existence of an extremely low ceiling to knowledge, which would drastically limit science and technology. Few scientists would go along with this possibility. But with an extremely low ceiling, every advancing technological civilization would meet it before long, thus making the anticipated technological marvels of the future impossible to achieve. Our technologies of a thousand years hence would then not appear to us as magic, as noted science fiction author Arthur Clarke once suggested they might. They would look rather similar to what we have now.

Alternatively, if the ceiling to a scientific understanding of the universe is high, or nonexistent, then the ceiling to technological applications from this understanding is also high or nonexistent. If infinite knowledge exists to be discovered, anyone with the necessary neurological equipment could continuously discover new aspects of nature and produce new technologies. It would all appear so magical, and the world we know would completely disappear along with us. Things would become so complicated that our descendants would soon have to engineer a new species of *Homo* to overcome our limitations. So what would we choose to be the reality: a low ceiling or a high ceiling to knowledge in the universe? Perhaps it would be better for the human race if we bump our questioning heads on a very low ceiling and live happily ever after.

Does all this have any relevance for SETI? It most certainly does. The amount of knowledge the universe offers for technological application is directly relevant to our thinking about where the aliens might be. If the ceiling to applicable science is very low, we may already be near a point beyond which nothing remains to be discovered that could lead to revolutionary new technologies. In this case, we will not be traveling very far from our earthly home. For SETI astronomers, a low ceiling is much better than a high one. A very low ceiling could mean that a propulsion system to cross the light-years in a routine manner is impossible. The first successful civilizations would then not have been able to spread throughout the galaxy, establishing bases in other planetary systems. A very low ceiling might even mean that no one is going beyond radio and lasers in their systems of communication, although the development of twisted light and twisted microwaves only fifty years after the first SETI projects shows that our current technologies in this area are still open to drastic changes. But if all aliens everywhere are stranded in their own planetary systems for lack of interstellar transport, and stuck with radio and lasers for their communications, then the SETI astronomers could be on the right track to contact other worlds. Alternatively, if the ceiling is high, then the aliens, or their artifacts, are more likely to be in the solar system. And their systems of communication might be as far removed from radio technology as radio is from smoke signals.

Yet perhaps the ceiling to knowledge is neither very low nor very high. Perhaps all successful civilizations will advance to meet a moderately high ceiling of knowledge, but it would have to be high enough to allow for interstellar transport. If that were the situation, then the solar system might have had visitors from different civilizations, especially during the past few hundred million years because they had detected the Earth's biomarker spectral lines.

They would all have reached the universal ceiling and would have applied the same available knowledge in the development of their technologies, including their spacecraft. In the same way in which we might expect all technologically intelligent species everywhere to have invented the wheel, we might also expect all highly advanced civilizations to have applied the control of inertia and gravity in their technology of spaceflight, if such control is possible.

A CEILING FOR ROBOTS

When we look at the rapid development of computer technology and artificial intelligence, and at our own emergence from upright apes a few million years ago, it seems as if human civilization might exist but briefly in the vast time frame of galactic history. The journey from the first four-legged land animals to the first humans took some 350 million years. Evolution is a slow process. But after just a couple of million years, since the first human species evolved, we have reached a point at which we might replace our human selves with a level of intelligence in robotic form far more powerful and durable than flesh and blood. The mechanisms that support intelligent robots may never match the wonderful complexity of life, but they might work more effectively. A car can travel many times faster than a horse, but the mechanisms that enable it to do so are thousands of times less sophisticated and complex than those that maintain the life of a horse. It would be the same with intelligent robots. They won't need our biological complexity to be considerably more capable than we are. It's therefore reasonable to wonder about a robotic age that might bring about problems that other worlds faced long ago. There are so many aspects of our civilization, including long-distance spaceflight, where intelligent robots could be employed, that their increasing use in the future seems inevitable. And a scenario where robots take over is not totally fanciful. It's something other worlds may have faced long ago. So what could we do?

Option one would be to enforce a worldwide restriction on the level of artificial intelligence. An impossible task. Some countries or organizations would break the ban for economic and political reasons.

Option two: Use genetic engineering and selective breeding to greatly improve the intelligence and capabilities of human beings so that biology keeps ahead of artificial intelligence. But raising intelligence in this way would be limited, while robots would not be limited by biology and what we could do with genetic engineering. The robots would improve much faster than we would.

Option three: Allow the robots to run the world and hope their programming, which we gave them until they started writing their own programs, would last and would make them tolerate us in the same planetary system.

A sci-fi story springs to mind from this line of thought. We see humans use their knowledge and skills in genetic engineering to keep ahead of the robots. A dramatic race follows. But the robots, being free from the limitations of biology and genetic engineering, are easily improving themselves with each new generation. Secretly they begin to reprogram themselves to deal with new opportunities for robotic life. In the end, the robots take over and *Homo sapiens* become extinct. But then the Earth is hit by a giant meteorite that destroys all sources of electrical power and makes the planet robot free. However, not every life-form perishes. A few intelligent rodents survive. Fast-forward fifty million years, and evolution has done a good job. We see the dominant species *Homo rodenticus* at work in their laboratories. And what are they doing? They're building their first computers.

Well, it's only a sci-fi story, and the arrival of a giant meteorite at just the right time is stretching credibility. But it did happen before. A giant meteorite did arrive at the right time sixty-five million years ago (although it was the wrong time for the dinosaurs). We wouldn't be here otherwise.

SPACE ARKS

Although nowadays confined to sci-fi dramas, interstellar space arks were a respectable subject in popular-science publications a few decades ago. They showed how future *Homo sapiens* could undertake slow interstellar journeys to the stars, taking plants and animals with them like a space-age Noah and maintaining a self-sustainable way of life with all the comforts of home. The drawback was that several generations would live and die on the ark before it reached a planet that could be colonized; otherwise, however, it didn't look impossible to some enthusiasts in the 1960s and '70s.

But now we can see that planets with biospheres could not be colonized by visitors with their own incompatible biology that they evolved on their home planet. This wouldn't be like space colonies in a future Dyson-O'Neill sphere, which could be serviced from the Earth and by other space colonies. It would be almost impossible to maintain the health of an ark's limited ecosystem for several hundred years in interstellar space, as experiments in the United States with sealed colonies of people, animals, and plants have shown. They failed within weeks. So, this looks like an impossible scenario for human beings—but not for intelligent robots who wouldn't need life support from an ecosystem. We can imagine robots drifting across the light-years in relatively small craft rather than arks, with the energy they needed to remain "alive" coming from the universe around them. The passing millennia wouldn't trouble them either. They could sleep the oblivion of electronic sleep for thousands of years and arrive in mint condition to explore those "brave new worlds" in ways that flesh-and-blood astronauts can never hope to do.

ANY RECENT VISITS?

For alien explorers, the past ten thousand years would have been by far the most interesting in the four-billion-year history of our planet. The birth of a new civilization must be a very rare event. If on average it takes a few billion years for life to develop civilization, we should be the only new civilization in the galaxy in the present epoch. And if intelligent ETs exist with the capability to travel here, they probably would not want to miss such an opportunity. Let's review the mainstream science that supports this conclusion.

1. It has taken four billion years from the origin of life on Earth to the evolution of technological intelligence.
2. The oldest stars that could have supported planetary systems with earthlike planets are about nine billion years old. Therefore, allowing for the development of planets and allowing four billion years for the evolution of tech-

nologists, our counterparts could have been active in the galaxy from about four billion years ago. Civilizations could then have emerged at intervals throughout galactic history, and we must be the latest to arrive.

3. There must be billions of planetary systems in our galaxy alone because so many have already been detected in nearby space.

4. The origin of life appears to be innate in the physics and chemistry of the universe. All the biochemical and molecular mechanisms of life on Earth have evolved on the basis of the universal physics and chemistry. The phenomenon of life will therefore come into play wherever the physical conditions allow it to form.

5. It therefore looks as if life is abundant in the universe. But what makes us cautious about assuming that our counterparts are also abundant is the fortuitous way in which technological intelligence evolved on Earth.

6. We note that only land animals with backbones, the vertebrates, have evolved big brains. And only a single vertebrate group, the primates, evolved the eyes and muscular coordination (through evolution in the trees) to give them the potential to evolve into technologists. And only a single group of upstanding primates, the hominids, diversified into a number of species that lived on the open savannas of Africa from five to two million years ago. Although some of these species probably had the potential to become civilized inhabitants of the world, only one evolutionary line led to *Homo sapiens*, while the others perished. Luckily for us, new environmental opportunities through climate change occurred at the right time to propel our ancestral line forward. If the widespread forests where many species of apes lived had not changed to open country at the right time, the evolutionary pressures on our ancestors would not have existed. All apes might then have stayed within the protection of the forests, as the ancestors of today's apes managed to do. Therefore, even with monkeys and apes well established in the forests of prehistory, there was nothing inevitable about the evolution of our ancestors.

Many mammals—and some birds—are intelligent, and there could be similar animals on other worlds, even though the builders of radio transmitters and flying saucers may not be among them. Go back to the bacteria and the bacterial-like archaea for a moment. These microbes were the dominant and only life-forms for the first two-thirds of Earth's history. Let's now rerun evolution on Earth. Countless creatures could have evolved without a sign of an upright ape, or even a monkey, or even a backbone, although given the evolution of trees, some life-form is going to make good use of tall trees. We can't even guess what candidates for technological status might have emerged, if any. So the path followed on Earth could be unique.

If predatory backboned fishes had not come ashore some 350 million years ago and started the evolution of all four-legged creatures, dry land would have been free for another aquatic life-form to take over the world. But from Earth history, the only possible candidate seems to be the octopoid life-form. Maybe the pressures and opportunities on dry land could have forced the octopoids to evolve a great diversity of species so that one line could have led to a technological species. Maybe not. Maybe a backbone of some sort and an upright stance are essential for anyone who is going to take over a planet. And maybe any visitors we have will be upright citizens of the galaxy, supported by some sort of backbone. For all we know, the Earth may have enthralled visitors for millions of years, or it may never have been seen by alien eyes. But if they came, those who made the interstellar spaceships possible would have been ETs who sat down quietly and discovered the secrets of nature.

Of course, we've had plenty of restless, aggressive people in human history, ready to fight their way to extinction, but not everyone has been like that. We've been lucky to have had enough quiet thinkers to maintain balance in a developing world civilization. And those who might go forth in the future to find "brave new worlds" will bear no resemblance to, say, eighteenth- and nineteenth-century *Homo sapiens* whose drive for land and wealth was overwhelming, no matter the cost to their fellow inhabitants. And so it might be with the advanced inhabitants of other worlds.

Another frequent question is, Would contact and communication be on the agenda of beings who had crossed interstellar space to study a planet overflowing with life? We communicate in a limited way with chimps, mainly in laboratory conditions. But the chimps are our nearest relatives, almost genetically identical to humans, as are the other apes. Communication with dolphins is another matter, though as intelligent mammals they are closely related to us. They are marine vertebrates that long ago evolved from land vertebrates not unlike scruffy bears. But do we worry that we haven't yet been able to communicate in a real sense with dolphins when most dolphins have bigger and more complex brains than the scientists who study them? That of course doesn't mean that they are smarter, and having evolved away the four feet that their ancestors once possessed, they're never going to get on the road to technological civilization. However, the ways in which dolphins respond to the cues we provide seem to show that they have learned more about our system of communication than we have about theirs. Anyway, we can see the problem of communications between related vertebrate life-forms in our own biosphere. That problem could be vastly greater with beings from other planets whose evolution and cultural development were quite different from ours. So when UFO contactees say that occupants of flying saucers have told them that they come from another world, they're right. It's called "Fairyland."

BACK ENGINEERING

Another fantasy used to persuade people that the aliens have arrived—and are intellectually close to us—is the "back engineering" of alien technology. Accounts from the back-engineering business are regularly found in the UFO literature and UFO films in which government employees—always ex-employees—talk about their back-engineering experiences. But there's a big problem here that the media has failed to address: back engineering of interstellar transport would be impossible. Anyone can see that our technology is based on our own science, which has mostly been discovered during the past few centuries—by us. Likewise, alien technology is going to be based on science that has been discovered during the past thousands or millions of years—by them. If we don't know the supporting science, we can't understand the technology on which it is based. And if we can't understand it, we won't be able to back engineer it.

Flying saucers, if they exist, would obviously be based on science unknown to us. If it were otherwise, we would be flying around in our own saucers. Think of the problem in terms of our history. There were good scientists and engineers in the eighteenth century, every bit as intelligent as those at work today. But had they been given a television or a computer, they could not have back engineered them because the science that supports these technological items was unknown at the time. Yet in this scenario, only a couple of hundred years of human scientific development separates our recent ancestors from today—not thousands or millions of years.

It makes sense, however, that if a flying saucer crashed anywhere on Earth today, the government would try to study it. But all the investigators would say the same thing: "We don't know how it works." Thus no one on Earth is going to duplicate UFO technology. The problem is that the fantasizers in this area of mythology are not shown up for what they are because journalists don't bring in the right people to ask the questions. A good reporter or anchor on a major network is ill equipped for the task. They don't know the right questions to ask that would drill holes in the back-engineering proponents' claims. Physicists are needed to question anyone who claims to have worked on a flying saucer propulsion system, and there are plenty of physicists who would love to sink their teeth into some back engineers and watch their fantasies die. Instead, the media uncritically promotes stories that are then picked up by other journalists who themselves are not equipped with the science to assess UFO reports. And so the stories grow to obscure what might be real phenomena. This is why we need rigorous science to construct testable hypotheses.

There's no hope if we think we can reconstruct crashed flying saucers, or go on trips around the Moon with visiting ufonauts, or believe in telepathic communications with beings from the stars. We've got to keep our scientific feet on planet Earth. It may be less exciting, but the payoff, if it comes, could be more exciting—and real. And if after years of research we find no evidence of an alien presence, we can all sleep soundly at night, knowing that no visiting aliens will abduct us and that all we have to worry about is the weakness of the human brain.

CHAPTER 6

TESTING TIME

It is now time to test the Grand Hypothesis and to look for the signatures of life in the various ways available to us. But before we stand a chance of shouting "Eureka!" certain conditions must have existed in the past.

First, a sufficient number of advanced civilizations must have evolved in our galaxy during the past four billion years for at least one of them to have discovered the solar system. There are of course billions of other galaxies like ours, so the opportunities for the evolution of high intelligence are unimaginable, but those other galaxies seem far too distant for our concern in SETI. Since there are a few billion sunlike stars in our own galaxy—potential sites for planetary systems—that is enough for the present.

Second, travel from planetary system to planetary system has to be possible, either by a highly advanced means of traveling through space or by the development of intelligent robots who wouldn't mind "sleeping" for thousands of years at a time while voyaging to other worlds.

Any of four basic lines of research could provide evidence that life is a universal phenomenon, though not all will test for the existence of both life and technological intelligence. For that, we have to find alien artifacts within the solar system or intelligent signals from other worlds.

The tests are as follows:

1. We could consider landing vehicles on Mars to investigate the planet's chemistry and to search for fossil microbes in exposed sites—or even live microbes under the planet's surface. Akin to this approach are plans to crash probes through the icy crust of Jupiter's moon Europa into the sea below. But there's a problem. Spacecraft

143

data indicate that Europa's icy shell may be between five and twenty miles thick and that the water beneath may be too acidic for life, at least for life as we know it. So the only way to find evidence may be to analyze material from cracks in the ice.

Saturn's moon Titan, with its dense atmosphere, is another candidate—but only for life that likes temperatures approaching absolute zero, if such a thing is possible. A probe has already penetrated this world's dense clouds to show us a landscape something like our own but formed by ultra-cold chemistry. The molecules that formed gases on Earth seem to have formed seas on Titan. And what is liquid water on Earth lies around Titan's landscape as solid rocks. This world is almost as cold as it can get, and any life here would have to be based on a different chemistry than ours. Most biologists say that life on Titan is not possible, but there are some who think it is.

2. Our neighboring world, Mars, seems the best bet, although microbes on Mars—alive or fossilized—might not answer our question about the universality of life. Large meteorites that hit Earth long ago could have splashed bacteria into space and then on to Mars. Alternatively, Martian microbes might have been splashed to Earth before life got started here. Gravity on Mars is a third of what it is here, so rocks containing microbes would splash more easily from Mars to Earth because of the lower escape velocity. If Mars had produced life before Earth did, it is likely that we would be Martians rather than Earthlings. Actually, without live Martians, we can't say whether Martian life is related to our life or whether it is independent. Martian fossils wouldn't provide information on the molecular biology of Martian life, which would be needed to confirm its independence from life on Earth.

3. Another test would be to image neighboring planetary systems (these are very rough images, but they are sufficient to show the positions of planets in their orbits). The spectra of life-marking molecules, such as ozone, could then be recorded to reveal the presence (or lack of presence) of highly evolved biospheres. If life-supporting planets exist in a proportion of nearby planetary systems, say, within thirty-five light-years (the region to be first studied by life-searching telescopes), then they will also exist in similar regions of our galaxy and in other galaxies throughout the universe. But this line of research could only confirm life as a universal phenomenon. The abundance or otherwise of technological intelligence would remain unknown. However, when astronomers find the life-marking spectral lines, the SETI astronomers are sure to search for intelligent signals. My guess is that they won't find any messages, but they might find spectral lines that indicate technological activity.

The European Space Agency's Darwin telescope and NASA's similar projects will search for the spectral lines of life in light reflected back through planetary atmospheres, so that the most easily detectable chemical contents of those atmospheres will be revealed.[1] We know the spectral signature of Earth, so astronomers could discover other Earths by searching for the main spectral lines in that signature. And it is this line of research, as astronomical technology improves, that might lead to spectral lines from technological activities being discovered, perhaps within Dyson-O'Neill spheres.

Of course, the immediate significance of the life-searching telescopes is that they answer the important question: How would other-world civilizations find a tiny planet such as the Earth, which orbits just one of thousands of stars in our part of the galaxy? Given that only one star in a thousand warms a planetary biosphere, we now know how that one star and its planets can be found before launching a single probe. For an advanced civilization, the Earth may have been relatively easy to discover after life here had produced enough oxygen to form a detectable ozone layer. If our counterparts are near enough, they could have found the solar system and gathered details about the Earth without leaving home. Whether or not they could have done more would have depended on their spaceflight capability.

4. Astronomical searches that can reach across the galaxy for radio or laser signals could confirm that intelligence is universal, which would obviously mean that life is universal. The near-fatal weakness of this approach is that it assumes that aliens will use our level of communications technology and will make the same guesses as we do about the best frequencies to use. The vast time frames in which planets and intelligent life evolve add to these problems. It would take a series of patient broadcasters stretching over a few billion years to detect any signals that might head this way in our time.

5. A final test would be the search for evidence of alien life or artifacts that might exist in the world around us or in nearby space. This type of test would involve local SETI as opposed to very long-distance astronomical SETI. A search like this might be on the fringe, but for the reasons already given, it looks to be a better bet than the search for broadcasts. It's also the most accessible and cheapest search opportunity—and the most neglected, mainly because it is linked to aspects of the UFO phenomena.

However, several astronomers, including Michael Papagiannis (see chapter 3), a professor of astronomy at Boston University, have searched for signs of alien artifacts in nearby areas of the solar system. Professor Papagiannis scanned the asteroid belt for any unnatural-looking asteroids. He said at the time: "The supposition that we are alone in the solar system is based essentially on the assumption that if others were here they would have made contact with us, or at least we would have become aware of their existence. Neither of these assumptions, however, is true, though it is possible that some of the thousands of UFO sightings might deserve some further consideration."[2]

The search for evidence of alien artifacts in the solar system has already been pursued at a high level in the science community. Papagiannis had a specific target, the spectral line of tritium, a product of nuclear fusion. The beauty of tritium as a target is that it has a half-life of only twelve years, so someone would have to be continually producing it for it to be present. Even so, focusing on a radioactive element involved in nuclear fusion may seem a bit chauvinistic; we're only guessing what technologies the aliens may have. Given that a lot more physics remains to be discovered, we should expect interstellar travelers to have something better and safer than nuclear fusion as their energy source.

Astronomers have also searched the Lagrangian points (see chapter 3) for the presence of alien probes. These so-called points are stable gravitational areas produced by the gravity of the Earth, Moon, and Sun. In theory, any incoming spacecraft could park in one of these locations and remain there indefinitely.

ALIEN ARCHAEOLOGY

Nowadays, there is a growing interest in searching for evidence of past visits to the Moon and Mars. As scientists at the Society for Planetary SETI Research would tell you, this pursuit is more promising than we might at first assume. Instead of evidence in the form of a signal in our immediate time, the search is on for evidence that arrived in the distant past and is still detectable. This is a potentially worthwhile line of research. No one would discover the solar system and the presence of the life-supporting Earth and not pay it a visit—if visiting were possible. And that would lead to visits to the Moon and Mars. The Moon especially would be an ideal base for the study and observation of Earth. This is obvious from what we know, yet those scientists who search for evidence in NASA's photography of the Moon and Mars are viewed almost as cranks. The problem is that this subject can become sensationalized among those people who lack scientific caution and who see weird and wonderful things on the Moon and Mars. Despite this problem, scientists at the Research Institute on Anomalous Phenomena (RIAP) in Ukraine and those at the Society for Planetary SETI Research have been active in this field of research.[3] Of course, attempts to find evidence of alien artifacts in the solar system give support to the best investigations of the UFO phenomena, which is something the science community would rather avoid. If there are alien probes orbiting in the solar system without our knowledge, then obviously there could be evidence anywhere: on the Moon or Mars or right here on Earth in the form of those pesky UFOs.

However, a brief look through the UFO literature will quickly reveal why most scientists avoid any unhealthy thoughts about UFOs. Let's look at the main speculations from ufology and see why—and why we couldn't test most of these ideas anyway.

1. Flying saucers come from a parallel universe that is so conveniently placed that aliens can jump from one universe to another. These universes would be rather large places, judging by the one we live in, so an alien would have to be a very precise jumper to arrive on Earth at the right time and place. The fact is that there is no evidence that parallel universes exist. At best, they're a theoretical concept, and we don't need to drag them in to explain the presence of aliens. We can do that well enough with just one universe. After all, the existence of other planets with intelligent life is no more unlikely than an Earth full of *Homo sapiens*. So let's leave parallel universes aside, which is where they might be anyway, if they existed. We can't go out into the field and investigate parallel universes, but we can perform forensic investigations of sites where seemingly reliable witnesses say UFOs have landed, as well as other scientific research that may either confirm the reality of UFOs or banish the subject to the realms of mythology.

2. One fantasy of ufology is that UFOs are created by human psychical powers. This is large-scale spoon bending and is a lot to expect from the human mind that can't even move tiny objects in controlled experiments. This idea is presented in the 1956 sci-fi film *The Forbidden Planet*, in which a long-extinct master race recklessly uses all its energy resources to build huge generators

to transform thoughts into physical reality. Thus the fear of alien monsters in the minds of the visiting astronauts created a massive monster that attacked them at night and stubbornly refused to die under intense fire from ray-guns. A good film, but not logically consistent. How did all the astronauts happen to fear the same monster? Logically, they should have created not one but several slightly different monsters, one for each astronaut's mental state.

3. Another interesting idea is that today's ufonauts are our descendants who've come from a distant future to see what their long-extinct ancestors are doing. The fact that the occupants of saucers are nearly always described as humanoids fits this sci-fi time travel fantasy. According to Einstein and other physicists, we can travel forward in time, providing we're traveling in near-speed-of-light transport. The prospect of going backward in time seems more unlikely, although there are a few on-the-fringe physicists who think that time travel both ways might be possible. If they are right, then the occupants of flying saucers might be our descendants. That there has been no provable contact between them and us would also fit this backward-time-travel idea. As all sci-fi enthusiasts know, time travelers must not interfere with the present.

4. The "skeptical detective" figures prominently in the UFO literature where investigators have shown that a much-publicized UFO case is a fantasy or hoax. The snag with this approach is that it requires a lot of time and effort to dismiss just one case, leaving the vast majority of cases unexamined by the skeptics. A scientific approach avoids this problem. We take the most credible cases—not necessarily true, but credible—and we use the consistent data to form hypotheses that are testable. In this way, we follow the traditional scientific method and save the "skeptical detectives" (and everyone else) a lot of time, while dodging all the psychical phenomena, Earth spirits, and parallel universes that litter ufology and demonstrate a major quirk of human reason— the attempt to explain the unknown in terms of the unknown. It's no use trying to explain UFOs, which we do not understand, by psychical phenomena and parallel universes of which we have no knowledge. Many people believe in the psychic realm, and no one can say definitively that it does not exist. But no scientific investigation has ever been able to detect its presence, so it shouldn't be used in any scientific explanation.

5. In dealing with the UFO phenomena, we have seen that it's only the relevant science that makes the subject worth any attention. However, I've yet to hear anyone provide scientific reasons for taking the UFO phenomena seriously. For example, the fact that planetary biospheres can be detected by their spectral

signatures has great significance for the extraterrestrial hypothesis of the UFO phenomena. Yet do any media commentators know about this? I doubt it. The point is that the science community can find ways of rejecting UFO reports—and plenty of UFO reports need rejecting—but scientists can't reject the synthesis of science that supports astronomical SETI. But having accepted this, they can't really get away from the fact that this same science also supports the extraterrestrial hypothesis of the UFO phenomena. They can only point out that much of what is published about UFOs is nonsense—which is true. Of course, this doesn't mean that the extraterrestrial hypothesis is correct. The explanation of UFOs as alien spacecraft may well be wrong. But the hypothesis that some reported UFOs are extraterrestrial technology is one that needs testing

THE DREADED MIN-MINS

One example of how to explain what appears to be beyond explanation comes from Australia. For many years, people were frightened by brilliant lights that followed their cars at night on dusty roads in the outback. The aborigines called these haunting lights the "min-mins," but drivers familiar with UFO mythology thought aliens were about to grab them. In every case, as the mysterious lights closed in, they abruptly disappeared.

Jack Pettigrew, a scientist at Queensland University, first met the min-mins when he was studying owls in the outback. He and some colleagues were driving at night when a light began to follow them. Being equipped for the outback, they were able to take the light's compass bearings. The light seemed quite near. They were therefore surprised, after driving about three miles, to find its position had changed by only one degree. This meant that although the light had seemed to follow the car's changes in direction, the position of the light itself had not moved. They were puzzled and got to work with a map. Their compass pointed to a town called Windorah, roughly 190 miles away on the other side of some hilly country.

At a later time, Pettigrew visited the town and discovered that a convoy of vehicles had passed through it at the time he and his colleagues had witnessed the min-min. He wondered if the light from those vehicles could have been bent over the hills to the location where they had seen the strange lights. In this kind of country, especially during winter's hot days and cool nights, a layer of cool air gets trapped under warmer layers, and that layer of cool air can act like an optical fiber, carrying light over long distances. But only on a later expedition, when camping with his group in the same region, did Pettigrew get a chance to test his theory. He drove his truck at night to some low ground about six miles away and pointed it toward the camp. He then got his colleagues on the two-way radio and switched the headlights on and off. Back at the camp, a min-min appeared and disappeared in time with the headlights.

It seems that this particular UFO phenomenon has been explained, especially as the min-mins appear only at night and usually in winter when a cool layer of air covers the land and can convey light over long distances. Physics and meteorology support this explanation, and min-mins can be produced experimentally. But it's understandable how the min-mins fooled people for so long. Imagine you're driving at night with a low Moon shining at you through a forest. You could drive a long way, and it won't change its position. If you didn't know what it was, you might imagine the aliens were after you.

REPEATABLE RESULTS

When scientists test any hypothesis, they want repeatable results. Different groups of researchers must repeat the same test and get the same results. No single result can stand alone. Some years ago, several distinguished physicists claimed to have discovered "cold fusion," which could have solved all our energy problems. It was a big claim from a reliable source, but was it true? Could other scientists get the same result? Teams of physicists in several countries got to work, but no one could repeat the results of the first study; therefore the reality of "cold fusion" was not accepted. To go from the established discipline of physics to ufology is quite a jump, but ufology, like physics, needs to provide repeatable results that the science community would have to seriously consider. But the ufology experts have not yet been able to provide this.

By way of illustrating this problem, let's look at a report on the Ubatuba

UFO. The twenty-five-page report describes samples made up of pure magnesium that witnesses said had fallen from a saucer that exploded over the sea near a beach in Ubatuba, São Paulo, Brazil, in 1957. The witnesses collected fragments of metal from shallow water, and these eventually reached Olavo Fontes, a medical doctor and UFO researcher, who had them analyzed. Chemists reported that the samples consisted of magnesium with a level of purity not possible to achieve at the time. On the basis of this limited analysis, plus a good helping of speculation about the exploding UFO, Fontes and others concluded that the fragments of metal probably had an extraterrestrial origin. Now, the scientific objection to this is not only that the evidence was not enough to confirm the major scientific hypothesis of our age, but also that it was a one-off event, which might—or might not—confirm an extraterrestrial origin for an exploding UFO. The snag is that UFOs don't explode often enough to provide regular samples of pure magnesium. If they did, scientists would eagerly await the next exploding UFO and the dispatch of the debris to their labs. In short, they would have an unlikely but repeatable experiment, and the extraterrestrial nature of the samples could be seriously considered. But it's not like that. The testable hypotheses in ufology are not exactly abundant, and the samples of substances that have been tested so far have been similar "one-offs." There has been no ongoing research.

WHERE SAUCERS HAVE LANDED

According to UFO organizations, thousands of saucers have landed during the past few decades. In 1975, Ted Phillips, the main pioneer in landing-site data, published the *Physical Trace Catalogue*, which lists 3,059 cases in 91 countries in which, according to reports, UFOs have left physical traces.[4] Phillips's case studies go back as far as the year 1490. Consequently, with so many alien visitors, we might expect to find a few calling cards in the form of chemical residues, as well as physical and biological evidence that provide some sort of pattern. UFOs in flight may not weigh much, if their propulsion system nullifies gravity, but their resting mass on the ground should leave detectable traces. Yet because no professional setup is ready for the next landing, weeks or months may pass before anyone arrives to investigate a reported landing—and then it's enthusiastic part-timers, not a team of dedicated scientists with the latest technology and know-how for a full forensic investigation. No wonder forensic reports on the chemistry, biology, and physics of UFO landings are hard to find.

It is not known whether there has been any significant data to be gathered. The authors of UFO articles and books are fond of describing all sorts of odd traces found at landing sites: unusual metals; fibers insoluble in sulfuric acid; strange crystals; tinted and phosphorescent liquids; dark, oily substances; silver powders; and so on. But such reports without the scientific details are useless. Some chemical analyses have been carried out, but, again, they are all one-offs. There is no pattern, nothing significant that might have been discovered if independent teams of scientists had been brought in to investigate, so we don't know what opportunities have been missed—if any.

Professional investigators, however, might eventually be able to predict what the findings might be from the next reported landing. We would then be engaged in science, in getting repeatable results from different teams of researchers. This is the way forward, but before we can make that move, we would have to persuade university researchers worldwide to participate in the heresy that alien artifacts might exist and be detectable. We would need a coordinating office—just one person with a phone and computer would do—and small teams of scientists doing their daily work but ready to rush to sites deemed credible enough for attention. They wouldn't be called on very often. Team members could be working at different universities, only meeting at sites at very short notice. After a few years, we could see a payoff. Would a significant pattern of evidence emerge? Ted Phillips certainly stresses the need to improve on current methods of data gathering. "Events must be well documented and sampled for laboratory analysis," he says. "There needs to be a central repository for evidence. Currently data is scattered among numerous investigators in many countries." In other words, the subject needs a rigorously run international scientific outfit.

Bill Chalker, a chemist and leading UFO investigator in Australia, has concentrated much of his research on trace cases. He believes this aspect of the UFO phenomena could be of great significance. "On the basis of the physical trace evidence, and the collective evidence contained within the whole spectrum of UFO evidence, I contend that a physical dimension to the UFO phenomenon has been substantiated. We now need to conclusively establish whether or not this physical evidence is consistent with a true alien reality."[5]

Chalker describes the dilemma professional scientists face if they dare to dabble in things ufological—even in their spare time. As a chemistry student at the University of New England, Chalker had problems analyzing traces from a reported UFO landing site in New South Wales, Australia. A colleague advised him to contact a Dr. Keith Bigg, who proved very helpful. Bigg's parting advice for Chalker was "Never admit that your interest is in UFOs. You're more likely to get co-operation in hunting witches."

DUTTON'S THEORY

The most developed and detailed theory to explain UFOs as alien probes is Roy Dutton's astronautical theory. It's a theory that could, if correct, give repeatable results in independent investigations, but it has never been tested in this way. Dutton used his professional skills as an aerospace engineer to develop the theory that the presence of an automated system of probes is monitoring our planet. But like all theories, this one will stand or fall by being tested.

Dutton began his research in 1967 after reading about the descriptions and flight characteristics of UFOs that people had seen in the Manchester area of England. He first analyzed the reports and plotted the locations of the events on a map. He found that the sites were within a straight north-south corridor only thirty-five miles wide. This suggested to him an aerial surveillance activity of some kind.

A series of UFO events a few years later in the same area resulted in a similar straight corridor, causing Dutton to wonder if this recorded data could relate to orbital activity above these areas of England. If so, it might be possible to plot reports of worldwide sightings and landings. The next step, then, would be to build up a global database of selected reports from all over the world of strange aerial craft.

Now, as everyone knows, the Earth continuously keeps turning. Consequently, a satellite orbit viewed from the Earth's turning surface will shift across the sky, following a spiral path in relation to the Earth's surface. Only geostationary satellites, which keep pace with our planet's rotation, remain fixed above a given location on the Earth. Any craft in orbit will therefore trace out a spiral path on the Earth below, and such a track can be plotted on a map. Dutton's discovery that the most credible UFO reports could be related to such spiral patterns indicated that UFO (SAC) events do have their origin in space.

At the time of this discovery, Dutton manually processed his findings, using a database of about 450 reports. Since then, his research has been fully computerized and now includes an additional 850 reports. Every event recorded in the database must have the date and the precise time of the event and its geographical coordinates; otherwise, it cannot be checked with the theory. But the times reported for UFO events initially presented a problem. Whoever or whatever might be delivering the SACs into the Earth's atmosphere, it's reasonable to assume that the control system would be governed by sidereal time, which is determined by the stars. But on Earth, we live by solar time (determined by Earth's rotation), and this happens to be about four minutes a day longer than time determined by the stars. Consequently, Dutton had to perform time conversions in order to check events. With the precise geographical coordinates and the dates and times of events, he can check them on the computer to see if they fit—to see if an event happened at or near the time that would relate it to one of the many ground tracks of the theory. Some reported events are spot-on. Some are near or just off a track, and some are nowhere near a track. But Dutton has found that about 70 percent of seemingly credible reports do relate to the theory's plotted tracks. In other words, the times and geographical coordinates given in the reports match the system of orbits predicted by the astronautical theory.

It is therefore possible to predict where in the sky a UFO might be present, though long periods of time might pass before anything could be detected by an astronomical monitoring system. This is because the theory predicts only what are called "virtual orbits" and "virtual entry points" from space—so named because craft would only sometimes be present. Such spacecraft might only seldom occupy the virtual orbits, but according to the theory, the orbits are there, part of the system used for delivering and retrieving those strange aerial craft. Whether we think this is possible is beside the point. The observational opportunities to test the astronautical theory—which is an obvious SETI theory—are clear, since the theory gives entry points in the sky and a range of different orbital paths. And it shows how these are related navigationally to the fixed stars and to sunrise and sunset. It's possible that advanced amateur astronomers could test the theory by searching for bursts of radiation at light, infrared, and microwave frequencies coming from Dutton's paths and entry/departure points in the sky, but the best astronomical technology may be needed to test the theory.

I have given here only a very brief and simplified account of the astronautical theory. Readers wishing to explore further can read Dutton's major technical papers, some of which can be downloaded from the web by searching for "T. R. Dutton." These papers show how the theory developed from the data with years of hard work and without preconceptions.

One aspect of Dutton's theory is that it incorporates the totally unexpected. It predicts that alien spacecraft, if present, will always be in retro orbit: traveling from east to west against the Earth's rotation (the opposite direction that most of our satellites travel). When Dutton was first searching for ground tracks, he assumed that the spacecraft delivering and retrieving the SACs into and from the atmosphere would orbit with Earth's rotation, but he was wrong. Then he considered that such craft might travel in retro orbits, against Earth's rotation. When he tested this idea, everything fell into place, but only after he tried an orbital period of 65.4 minutes, a time suggested by the processing. That the data indicated this solution was surprising because a speed of about twenty-five thousand miles an hour would be required to keep craft in such retro orbits. This would demand massive amounts of energy, something our present technology couldn't provide.

CONVENIENT FREQUENCIES

While the astronautical theory shows astronomers where to look, certain UFO data suggests that SETI astronomers are already near the right wavelength to check Dutton's theory. Hundreds of witnesses have reported close encounters with SACs, and many have described injuries, especially burns, which seem to indicate the presence of microwave radiation. Of course, radiation in other frequencies (from gamma rays to long radio waves) might be present, but a high proportion of the biological effects that close-encounter witnesses have reported over many years can be explained (though no more than that) by exposure to microwave radiation from 1GHz to 3GHz.

Apart from burns, numerous close-encounter witnesses over many years have reported being unable to move. A UFO study by the British government mentions this "temporary paralysis of humans until the nearby object moved away." It goes on to say, "One explanation is that either constant or pulsed microwaves, coupling into the nervous system, may cause this affect."

James McCampbell, an engineer who at one time worked for NASA, is the pioneer researcher in this specific area, although other scientists have published papers describing research on the biological effects of microwave radiation at these frequencies. In the 1970s, McCampbell was already explaining how microwaves at about 3 GHz could impact on the biology of nerve fibers to produce this temporary paralysis.[6] So, a connection between UFOs and microwaves was seriously considered over thirty years ago—at least in theory. Anyway, the point for us is that the frequency band of 1GHz to 3GHz conveniently includes the "water hole" (see chapter 3) band of frequencies (from hydrogen at 1.42 GHz at one end and the hydroxyl radical at 1.72 GHz at the other) that continues to be the main region of the spectrum scanned by SETI radio astronomers. So if SACs radiated the same range of microwaves in space as they may do on the ground, there is an obvious opportunity for SETI radio astronomers to perform a long-awaited test of Dutton's theory.

Optical astronomers, too, could scan the frequencies of light and infrared radiation that might be coming from the orbits and entry points predicted by the theory. The Kingsland Observatory in Ireland—mentioned in chapter 2—built and operated by astronomer Eamonn Ansbro, has checked the theory with automatically controlled telescopes, looking for evidence in visible light and the near infrared parts of the spectrum. Ansbro is a designer of optical instruments and has built special equipment to record fast-moving objects in Earth orbits and in the atmosphere. He is interested in recording spectra of bright moving targets, which requires both all-sky tracking and spectrographs to record spectra showing the targets' nature. His all-sky cameras can identify targets and trigger a video system to capture on tape the spectra of any passing probes.

A few radio astronomers, perhaps members of the SETI League, might also like to join the search for probes after considering the possible link between the microwaves in "close encounters" and the fact that astronomical SETI is already equipped to monitor such frequencies. They wouldn't have to change a thing—just monitor certain locations in the sky. In fact, they could continue to search for alien messages at the same time. Dr. Paul Shuch, director of the SETI League, has said, "We encourage SETI League members to scan the Water Hole, yes, but also any other interesting frequencies which strike their fancy. That way, maybe *someone* will guess right."[7] When he said this, Shuch was thinking about the detection of signals from other planetary systems, but I guess he would approve if some of his members broke ranks and began the search for evidence of orbiting probes.

This opportunity would also be open to the major development in SETI in recent years: the SETI at Home program, in which *a few million participants* use their home computers to search for signs of intelligence on other worlds. Those participating download a special program to their computers to search through observational data provided by the giant Arecibo telescope in Puerto Rico, the world's largest radio dish. The project has been scanning a narrow microwave band centered on 1.42 GHz, the frequency of neutral hydrogen and the most common frequency, because hydrogen is by far the most common element in the universe. It is therefore a major source of astronomical data and consequently considered by SETI astronomers to be the frequency that other worlds would use to contact their neighbors. But some SETI at Home participants might like to use their computers to search for the microwave signatures of alien probes. As data from the Arecibo radio telescope come with precise times and coordinates, any "hits" could be compared with the virtual orbits of Dutton's theory. So, ideally, and given enough time, the whole system of Dutton's orbits might be duplicated from new observational data, though the Arecibo dish does not cover the whole sky and some observational data would have to come from elsewhere.

It's a very long shot, of course, but perhaps no longer than searching for messages from the stars transmitted by radio technology, which, by happy coincidence, we just happen to have available. And by contrast, the new search would be based on reported data gathered and studied over many years. Some data will be false, but the large number of reports that show consistency in their details tends to counter this problem.

MICROWAVES AND CLOSE ENCOUNTERS

Let's briefly return to the link between microwaves and close encounters, because it has to be a credible link to justify scanning the skies for specific microwave frequencies. First, there are hundreds of witnesses who have reported a variety of physiological effects, mainly excessive heat and burns, including "sunburn" at night and through clothing. People in cars with the windows down have reported the top halves of their bodies getting burned while the metal of car doors protected the rest of their bodies, presumably by reflecting the microwaves. The British government's secret four-year study of UFOs (2000–2004) describes one case: "In one event the witness, a policeman, suffered arm-burns through a long sleeved uniform. But his body was otherwise protected by the car door, behind which he was standing." It points out that infrared or ultraviolet radiation could not have been responsible, that the burns must have been due to microwave penetration through the policeman's clothing. Could that radiation have come from a hot ball of plasma, or from something else?

There are numerous cases in which metallic objects have become hot, such as rings that became too hot to wear. These close-encounter witnesses have not only reported what they've seen but have sometimes had the effects of their experiences to show others, which makes their reports more credible than just a UFO report. Of course, this doesn't make their data acceptable, but it does make it usable—to form testable hypotheses. But it's the results from testing that count.

These days, we know a lot about the various biological effects of microwaves because of the development of microwave ovens, although, as previously noted, James McCampbell linked the injuries and symptoms of close-encounter witnesses to a microwave frequency range as early as the 1970s. The main experience reported is one of excessive heat, and some witnesses have reported the heat as unbearable. There is a report of an aircrew that became so overheated by a nearby UFO that they bailed out. How reliable such reports are is difficult to say. They usually can't be followed up to determine the truth, so some are bound to be misleading or manufactured.

But we needn't worry too much about this because there are so many reports showing an intriguing consistency. Either something is out there or all the witnesses have been reading the same UFO literature—we don't know what the truth is. So unless someone has some dead aliens stored in their deep freeze or the remains of a crashed UFO in their backyard, we have to use science in the search for answers. It may all lead nowhere or to the confirmation of the Grand Hypothesis.

An event reported in 1957 is worth mentioning here because it seems to be the best official evidence available. A US Air Force RB-47 monitored a microwave frequency of 3GHz when, it was claimed, the crew was in visual contact with a UFO for one and a half hours. The RB-47 was on a surveillance training exercise with the best electronic monitoring systems of the time, when it encountered a UFO at high altitude. Two official reports describe how the UFO tracked the RB-47.[8] Although this case is unusual in the amount of information recorded, there are other cases of aircrafts' onboard electronic systems being disrupted—possibly by microwave radiation—while crews were in visual contact with structured craft. In some cases, these visual observations have been confirmed by ground radar.

Dr. Richard Haines, a retired NASA scientist, and Dominique Weinstein, technical adviser to the National Investigations Committee on Aerial Phenomena, compiled a catalog of sixty-four cases where onboard electrical systems, including radios, compasses, and direction finders, were temporarily affected when aircrews were in visual contact with UFOs. Dr. Haines also compiled a catalog of more than three thousand aircrew reports, and Weinstein has assembled a catalog of 1,305 cases, some involving both air and ground radar and sightings from more than one aircraft. Most of the reports involve military aircraft and commercial airliners, though some are secondhand and not exactly reliable. Other events listed may be reporting atmospheric plasmas, but there are reports that describe structured craft. Again, we are dealing here with "reports" rather than data from testing, but these reports are about the best we can get. And they seem to indicate that microwave radiation is associated with UFOs in flight.

Ideally, we would have enough data to provide a distribution curve that shows the highest peak of microwave radiation from UFOs. Radio astronomers could then focus on that peak frequency. But even though such precision about peak radiation is not possible, SETI astronomers who today scan the famous water hole microwave band are already set up for a new line in SETI research. Of course, the UFO propulsion system must radiate microwaves in orbit, as many reports suggest it does when it is near the ground—and when UFOs are buzzing aircraft. On this subject, the catalogs compiled by Haines on the electromagnetic interference of aircraft systems when crews have been in close visual contact with UFOs, are the main source of data. But here there's a need to determine the frequencies and intensities of microwave radiation that would produce the reported instrumental malfunctions. This is a job for electronics specialists, but some idea of the range and intensities of electromagnetic frequencies coming from those controversial craft could be important for future research.

Some researchers have pointed out that the hazy luminosity reported and captured on daytime videos of UFOs could be produced by microwave radiation raising the energy state of the molecules of air (nitrogen and oxygen), causing energy in the form of light to be radiated. But there seem to be no research papers by physicists on this phenomenon. Although such research would not provide any acceptable proof that those strange aerial craft have a physical reality, it might eventually contribute to this. If different lines of research point in the same direction, the pressure for appropriate scientific involvement will increase. SETI scientists may say—will say—that the data on microwaves and UFOs is not reliable. And that is true. But also unreliable is the assumption that the inhabitants of Planet X are busily broadcasting across the galaxy. Both radio and optical astronomers have to take a risk to move forward with SETI, and so does every other investigator.

LIFE MARKERS AMONG THE STARS

On a clear night, away from city lights, we can see beyond the nearest stars to those stars so distant and numerous that they look like a band of luminous mist across the sky. We are looking into our galaxy, the Milky Way. From where we stand, nearby stars may look close enough to each other for space travelers to easily cross the void, but the stars are separated by light-years and only appear close to each other in our line of sight. Astronomers tell us that only a small proportion is like the Sun in size and temperature, although that "small proportion" amounts to a few billion in our galaxy alone.

The sunlike stars in our neighborhood reveal their status by their optical spectra, but we don't yet know which have earthlike planets in orbit. These would be between half and twice the mass of the Earth. Planets with less than half the mass of Earth would probably lack sufficient gravity to hold their atmospheres. It could be that most single sunlike stars have an earthlike planet in a suitable orbit for life; only the new "life-finder" telescopes may give us the answer. But before anyone can find the life markers—the spectral signatures of life—from other planetary systems, it's necessary to know which stars in our neighborhood have planets in orbit. Finding planets that can support life will be the second phase of the search. And evidence of life is one thing; evidence of intelligent life is another.

SETI scientists realize that only planets with a high metal content are likely to support technological civilizations. It follows therefore that as planets form from the same material as their suns—from stellar discs—technological worlds would have to have suns with a high metal content. This line of thinking predated the discoveries in recent years of other planetary systems, but now astronomers have made the remarkable discovery that stars with high metal content are the ones that are more likely to have planets, rather than stars with a low metal content. It seems that metals in a stellar disc help particles stick together to form larger and larger chunks, until the force of gravity kicks in strongly and helps to combine matter into proto-planets and planet-sized bodies.

Astronomers know the compositions of stars from the lines in their spectra, which show the elements and molecules present. In a research project conducted by Debra Fischer of California University and Jeff Valenti of the Space Telescope Science Institute in Baltimore, the spectra of 754 stars were compared with the Sun's for their iron content. Fischer and Valenti divided the stars into three groups according to the amount of iron present. The lowest group of 29 stars had no planets that astronomers have so far detected. The second group of 350 stars had 15 planets already detected. And in the top group of 388 iron-rich stars, there were 46 planets so far detected. This linking of iron and other metals that stars possess with the presence of planets is a significant advance. It means that astronomers searching for the presence of Earthlike planets, using the new "life-searching" telescopes, have the best chance of finding planets with life by targeting stars with a high metal content. SETI astronomers searching for signals will do the same, because without metals the emergence of alien broadcasters would be impossible. As Debra Fischer says, "We now know that stars which are abundant in heavy metals are five times more likely to harbor orbiting planets than are stars deficient in metals. If you look at the metal-rich stars, 20 percent have planets. That's stunning." Fischer's colleague, Jeff Valenti, adds, "The metals are the seeds from which planets form."[9] In other words, the metals in stellar discs can clump together more readily than nonmetals, thereby providing the initial chunks of matter that will gravitationally accumulate other material and grow into real planets.

PLANETS FOR LIFE

In 1995, the first planet orbiting another star was discovered, and since then planet hunting has boomed in the world of astronomy. Hundreds of planets have been found, and the total keeps rising. This is the first step toward discovering evidence of life on faraway worlds, though the discovery of any type of planets and their orbits within their systems will eventually show if the layout of our solar system is common or highly unusual. So far there have been some big surprises. Many giant planets, up to about ten times the mass of Jupiter, have been discovered orbiting very close to their suns, some closer than Mercury is to the Sun. This is contrary to old theories of planetary formation that predicted that the solar system would be typical, with giant planets orbiting in the outer part of a system and rocky planets in the inner part. But the unexpected is always turning up in science, and this is a good example. Nevertheless, astronomers do expect to find plenty of solar systems that have a similar pattern to the solar system, with rocky planets (like Mercury, Venus, Earth, and Mars) in the inner regions while gas giants (like Jupiter, Saturn, Uranus, and Neptune) orbit in the outer regions.

It was no surprise that massive planets were the first to be discovered, since the main technique depends on the gravitational effect of a planet moving its sun around a common center of gravity. The greater a planet's mass and the closer it is to its sun, the easier it is to detect by its gravitational effect. But what has surprised astronomers is that in many systems massive planets are so close to their suns. By the way, the amount of a sun's movements (due to the gravitational effect of a close planet) is determined by the shifts in frequencies coming from it. As a sun is moving away from the Earth, the wavelengths of its light get stretched out. As it moves toward us, its wavelengths become shorter.

The second method of detecting other worlds is to look for the transit of a planet across the face of its sun. Such events affect the amount of light from the Sun—in a very small way but enough for astronomers to extract detailed information. Many planets have been discovered, but they are not a representative lot since they are the ones astronomers would expect to detect by the methods used. They are all too massive to be Earthlike planets. But new astronomical telescopes that can detect Earthlike planets (and the presence of life) will be launched in the decades ahead. They will be the most sensitive astronomical telescopes ever built and over a period of many years will scan at least a hundred thousand stars in our neighborhood of the galaxy for planetary systems.

What is astonishing is the number of planet-hunting projects that are already observing or are at least in the planning stage. NASA and the European Space Agency are the major players, but there are twenty space telescopes and an amazing seventy-nine ground-based projects, though some of these are just confirming planetary discoveries already made. Even advanced amateurs are having a go in this new realm of astronomy. That there are so many planets out there is a great boost for SETI—especially for local SETI, if interstellar space is crossable.

The ultimate stage in planet hunting will be the direct imaging of planetary systems and individual worlds. At one time, this looked like an impossible task because in optical light a star would outshine a planet a million times. But this number drops to about a thousand times if infrared radiation is used. This fact plus a lot of advanced astronomical technology will enable astronomers to make direct images of planets orbiting other suns. Spectroscopic studies will then find those planets radiating in their spectral lines the fingerprints of life—even spectral lines that reveal the presence of technology in action. This will be the most exciting part of the planet-hunting business. Specialists in spectroscopy will spend years interpreting the spectra from other planetary systems and suspected Dyson-O'Neill spheres. What they find may revolutionize our view of ourselves and the universe we inhabit.

The movement toward this target can be seen in the development of special space telescopes. Both NASA and the European Space Agency are preparing to build space telescopes with four or five separate dishes that will together provide the "seeing" power of a telescope about three hundred feet in diameter. With such "life-finding" telescopes, astronomers will image and study our neighboring Earthlike planets—*neighboring* means within about thirty-five light-years. Astronomers and biologists will be looking for biomarker lines in the mid-infrared part of the spectrum, the same lines that alien astronomers could have observed coming from Earth, if they had been looking in our direction at any time during the past few hundred million years.

Some skeptics say we shouldn't expect to find the biomarkers we already know about because life on other worlds may function in different ways. Perhaps, but we should still look for the markers that we know life does produce. The chemistry of life on Earth has been so durable and successful for

3.5 billion years that it could be equally so on other worlds. The physics and chemistry of life is horrendously complex and only partially understood, but it works amazingly well. It's the way it is because of the way it has evolved within the limitations set by the universal constants that we reviewed earlier.

Totally different chemical and physical systems can be envisaged for life, but only in a superficial sci-fi way. That doesn't mean that many variations of our own chemical design do not exist, but those variations would still be "life as we know it" based on carbon and water. The search will be on for the spectral lines of ozone, carbon dioxide, and water. Carbon dioxide has a prominent line at 15 microns in the infrared, and the 6.3 micron line of water would come from water vapor in atmospheres—and water vapor could exist only if there is liquid water on a planet's surface. Oxygen, however, lacks adequate spectral lines to be detected by a telescope such as Darwin, and the 7.7 micron line of methane (a continuous product of life) would not be detectable at the level at which it exists in the Earth's atmosphere. Nevertheless, although oxygen cannot be detected directly, it can be detected indirectly through the presence of ozone, which is composed of three oxygen atoms. Thus the detection of the prominent absorption line of ozone will be a major objective.

A member of the Darwin team told me that "the spectroscopic absorption band of ozone from 9.5 to 9.7 microns wide results from the spread of the prominent spectral line of ozone at the wavelength of 9.6 microns. It's in the mid-infrared part of the electromagnetic spectrum, and the absorption band could come only from ozone layers similar to our own." This means that an ozone layer could not be present in an atmosphere low in oxygen. It would need photosynthesis of some kind to evolve, because almost all energy for life on Earth comes from the Sun via this process. It's reasonable to conclude that once life gets started on a planet, it would not neglect the constant availability of energy that its star provides, and it would release oxygen that might eventually form an ozone layer.

Geologists and astronomers tell us that there was only a trace of oxygen in the Earth's primordial atmosphere. This is important because life could not have formed where free oxygen was present. The chemical stages that led to life would have been oxidized away before the first self-replicating systems formed. But the oxygen in our atmosphere was largely due to photosynthesizing microbes—bacteria and archaea—microbes that are genetically distinct from bacteria. Later, plants provided oxygen and food for other life by turning light energy from the Sun into energy-rich molecules.

What is relevant here for SETI is that the Earth has had a detectable ozone layer for only about one-eighth of life's history. A large proportion of planets in other systems may support plenty of microbes but have no detectable ozone layer yet because the geological evidence indicates that the buildup of atmospheric oxygen took a very long time. The fossil record shows that only from about 1.5 billion years ago did advanced single-celled plants add to the process begun by the photosynthesizing bacteria and archaea. Multicellular green plants evolved much later.

One recent discovery adds to the existing evidence that photosynthesis should be universal. Soon after the dawn of life, microbes must have divided genetically into two quite separate groups: bacteria and archaea. And species within both groups independently evolved photosynthesis, though in each group this process operates differently. Throughout life's history, most of the oxygen released by life disappeared from the atmosphere into the oxidation of rocks on land and in the oceans. Oxygen is a very reactive gas, and something like 50 percent by weight of surface rocks consists of oxygen combined with other elements. And it has been estimated that without continuous photosynthesis by plants and microorganisms, all oxygen would disappear from our atmosphere within four million years.

Photosynthesis looks like a universal process because life on other worlds is not likely to ignore a free lunch. It's chemically possible to evolve photosynthesis, so, given the opportunity, life will do so to trap that free energy. Even on Earth the systems of photosynthesis are not all chemically identical, the biggest difference being between the two microbial groups—bacteria and archaea. But we need not worry about such differences on other worlds as long as oxygen is produced as a by-product. Given that it is, it looks as if the detection of the prominent ozone line could only come from a planet where life has been active for a long time.

Geologists believe there could have been enough oxygen in the atmosphere to establish an ozone layer by around six hundred million years ago. However, the fossil record most definitely confirms that by 350 million years ago there was a substantial ozone layer in place because it protected the newly evolving land life (which we can date) from lethal solar radiation. At that time, a certain group of fishes had evolved simple "lungs" to live partly out of water, probably starting on the mud flats of estuaries where later they evolved into the first amphibians. That evolutionary step from water to land required the protection of an ozone layer. The subsequent colonization of all habitats on dry land confirms that this step remained in place, which supports the fact that the presence of the prominent spectral line from planetary ozone layers would confirm the presence of life, though not, of course, technologically intelligent life, which the SETI astronomers search for.

It seems very convenient that certain spectral lines, which depend on the nature of the universal physics and chemistry, offer a way for intelligent life to find other life on other worlds. And the detection of that spectral information, rather than the detection of alien broadcasts, could be the way that advanced civilizations find life on other planets before launching exploratory probes. We need not worry, as some have, that our own broadcasts have betrayed our presence to aliens with an unwelcome interest in "strange new worlds." The spectral lines from our atmosphere gave the game away at least 350 million years ago, and the aliens haven't eaten us yet—and they never will.

SETI AT HOME

The idea came to David Gedye and Craig Kasnoff at a Christmas party in Seattle in 1994. SETI scientists need a lot of computing power to analyze the data their telescopes receive, and the most powerful computers are beyond their budget. So what was the solution? Perhaps, instead of one very powerful computer, they could use many small computers, like those operated by people in their homes. And so SETI at Home was born. The way it has developed in recent years is astonishing and unique in science: five million people around the world use their personal computers to search for intelligent signals from across the galaxy. And as the power and number of home computers grow, so does the capacity of the SETI at Home project, which already has a greater processing capacity than the world's most powerful computer. "The computers are doing the listening and the more computer power you have, the better job you can do,"[10] says Dan Werthimer at the University of California–Berkeley, where SETI at Home is based.

The SERENDIP (Search for Extraterrestrial Radio Emissions from Nearby Developed Intelligent Populations) program, also at University of California, uses a supercomputer receiver that can be plugged into a radio telescope without interfering with the observations in progress. The resulting data is processed for any intelligent signals that aren't originating from Earth. The first SERENDIP receiver was built in the 1980s, and the current model has been piggybacking on the Arecibo telescope in Puerto Rico, the largest radio dish in the world.

The biggest advantage of piggybacking is that you have the world's largest telescope "24 hours a day all year round," says Werthimer. "Most astronomers are lucky to get a day or two a year" on the telescope to carry out their research.[11] SERENDIP also operates in Australia and Italy, recording observational data for the SETI at Home program. Like most other projects in astronomical SETI, this program scans a narrow frequency band centered on the 1.42 GHz frequency of neutral hydrogen. Admittedly, this frequency is a major source of astronomical information, so it might be the choice of broadcasters on other worlds who want to tell their story to the galaxy. SETI at Home is another "long shot," but it carries millions of people into areas of science that help to clarify the human situation, and it might discover the unexpected.

PROJECT PHOENIX

Project Phoenix started as NASA's SETI program, but "down to Earth" politicians in Washington stopped its funding. The SETI Institute in California and private enterprise then took over the program. It has scanned more than a thousand sunlike stars within 200 light-years and is now using the Allen Telescope Array, a revolutionary development partly supported by Paul Allen, once a partner of Bill Gates at Microsoft. The finished telescope has 350 dishes, each one 6.1 meters (20 feet) across. So it's revolutionary technology, like many telescopes in the world today.

The SETI Institute has done more than search for aliens. It has brought the public into the science of life and the universe, although no one at the SETI Institute would dare to entertain the diabolical idea that the scientific work on the UFO phenomena is also part of SETI. In reckless moments some staff have admitted that there could be visiting probes in the solar system, but only as long as they're not UFOs, or even worse, flying saucers. The probes must keep their distance and preferably be very ancient, orbiting somewhere near Jupiter or farther out. They can't come near the Earth and be seen by people. Such a suggestion is scientific heresy of the highest order and quite impossible. And they could never have landed on the Moon or Mars, even though any probes to arrive will have come to study the whole solar system.

SETI SEES THE LIGHT

We know that when SETI got going in the 1960s, radio telescopes were conveniently available. So the first astronomers to search for our cosmic neighbors looked for messages sent on the frequencies that our radio telescopes were built to receive. And physicists Philip Morrison and Giuseppe Cocconi had already suggested that the prime target should be the frequency of neutral hydrogen at 1.42 GHz, since hydrogen is the most abundant element in the universe and its frequency is a major carrier of astronomical data. Astronomers on other worlds would be observing this frequency, so it seemed an obvious one for cosmic communications. SETI astronomers accepted this and tuned their receivers accordingly. More than fifty different searches were begun and major programs are still active.

The idea of looking for signals in light frequencies didn't catch on until later because when SETI began in the 1960s laser technology was in its infancy. I remember people looking around for something to do with the laser. It seems very odd now, but it became known as the invention without an application. But that didn't last long. Lasers soon had major applications in our technologies. However, it was initially thought that in the frequencies of light the parent sun would greatly outshine any signals that communicating aliens might send our way. "Not so," said the SETI optical astronomers. As

Monte Ross, one of the pioneers in the subject, explained: "With transmissions in brief bursts, each pulse could easily be a thousand times as bright as any nearby star in the receiving telescope's field of view. The shorter the pulse the less background light there is per pulse to compete with the signal."[12] And optical SETI offers other advantages over observing microwaves with radio telescopes. The microwave band for SETI radio astronomers occupies a very tiny part of the electromagnetic spectrum compared with that offered by visible light and the near-infrared. There is also a noise problem with microwave frequencies—from terrestrial sources and even from the temperature of the receivers being used.

Scientists at the University of California–Berkeley have been running two optical SETI programs. Geoff Marcy, the pioneer planet discoverer, has looked for laser signals from a thousand stars, searching spectra for extremely sharp spectral lines. The other program at Berkeley has focused on 2,500 stars similar to the Sun, though some are much brighter and some much dimmer. With a telescope specially built in 1997, astronomers have searched for pulses of laser light that might last only a billionth of a second. A few globular clusters of stars and some distant galaxies have been scanned in addition to neighboring stars, but not a flicker of an intelligent signal has been found.

The University of California–Berkeley is at the heart of astronomical SETI these days. Its famous Lick Observatory also began optical SETI in 2000 and by January 2004 had scanned 3,999 stars, giving each star ten minutes—not much for diligent aliens broadcasting hopefully for hundreds of years. The big disadvantage with laser SETI, of course, is that the aliens would have to beam their signals directly at a target, which I suppose they might do after detecting the spectral lines of life on a planet—or, much better, technologically produced spectral lines.

PHOTONSTAR

We can see from the problems involved that astronomical SETI may eventually give up searching for messages or any thoughts of sending our own messages into the cosmos. But SETI astronomers are going to be essential in future searches for evidence of alien life and intelligence. The PhotonStar project is a case where the search for intelligent signals could give way to searching for unusual light phenomena. PhotonStar is yet to start, due to the availability of certain technology, but it would use Global Positioning Satellites, the Internet, and thousands of small optical telescopes owned by enthusiastic amateur astronomers throughout the world. PhotonStar is like SETI at Home, except that each participant must have a telescope. The Global Positioning Satellites would determine the precise positions of all these telescopes in the project, wherever they were in the world. In this way all observations could be integrated, thereby creating the equivalent of one very large optical telescope. As Stuart Kingsley has summarized, the Internet would be used to gather data from each individual astronomer at a central computer for correlation. Each participant would have a laser receiver programmed to identify interesting signals, and all the telescopes would simultaneously be focused on the same star. These automated telescopes could be on line night and day.

However, PhotonStar would face the same problem that radio astronomers have faced in SETI: it assumes that a long series of civilizations in the history of the galaxy will transmit signals that they know may never be received. Moreover, these signals would have to be transmitted with technology that is compatible with what we are currently using. These are bold assumptions dependent on an abundance of alien worlds that want to tell the galaxy about their existence. At present no one knows the right target stars among the multitude of targets available. But after the planet-imaging spacecraft provide data, it might be possible to detect evidence of distant technological activity through spectral lines that could only come from such activity. But for the present, there is a more direct approach that some optical astronomers could try. Optical SETI could test Dutton's astronautical theory, which provides a complex guide on where to look for evidence. The theory may be wrong, of course, but anyone can check it out before adjusting their telescope for the search. So far only one astronomer has done so: Eamonn Ansbro at the Kingsland Observatory in Ireland, who has made optical observations to check it.

THE LESSON FROM TWISTED LIGHT

Having considered astronomical SETI and its progress in developing sophisticated technology to search for intelligent signals, we had better close on its main problem. We know it is reckless to try to second-guess the aliens, but those in astronomical SETI have based their careers on doing just that. They might be lucky, but it's obvious from history that we can't even guess what technologies our own civilization will be using in the near future. Yet for almost fifty years, SETI astronomers have based their research programs on what they assume civilizations many thousands or millions of years ahead of us will use to communicate across the light-years. They've assumed that advanced civilizations will be using our communications technologies, though in a more powerful form—like North American Indians expecting smoke signals but from bigger fires. The hard fact is that if other worlds aren't using our present-day technologies, we don't have a hope of detecting their signals.

SETI astronomers do acknowledge that their line of research is a long shot, but they're encouraged to think success is possible because of the vast number of sunlike stars that could support planetary civilizations. They hope that at least someone is at about our technological level. The other straw to clutch hold of is the possibility that frequencies may be detected—by optical or radio telescopes—which could not come from natural sources. That's a more substantial straw: that the activities of alien technologies could be producing by-products that might be detectable by their spectra, which could be detected by our current equipment. We wouldn't have to possess the technologies of super aliens to receive such data. That's some relief because this is otherwise an impossible problem to solve, as a recent development in laser and microwave communications demonstrates. After only fifty years of SETI, twisted light technology is being developed on Earth—specifically at the University of Glasgow. And it reinforces the point that you can't second-guess the aliens. It's a development that offers a better way of transmitting information. By twisting your transmitted signal, whether it is in laser or microwave form, more information can be carried than with a conventional signal of the type that both radio and optical SETI astronomers search for—and which we use daily in our communications. Twisted light produces a tighter beam, making it more energy efficient, and the more you twist it, the more information it will carry. Professor Miles Padgett and his team at the University of Glasgow have given light eight twists. When this is increased to sixty-four twists the beam will carry six times more data than an ordinary laser signal. Twisted light or twisted microwaves should therefore provide a better way of transmitting data to specific targets than any system previously in use. It's ideal for transmitting information through the atmosphere and through space: from building in cities or from the ground to satellites. Microwave and laser communications may never be the same again once the new technology is developed.

The lesson for SETI is obvious—and it's not that the aliens will be using twisted light, which they may have given up a million years ago. It is that after only fifty years a single development in our own communications technology has made past SETI searches look naive, since they've been searching for something that we ourselves will not be using in the near future. Of course, to think that the aliens might be targeting the solar system with twisted microwaves or twisted light would simply be another case of chauvinism, and we shouldn't think like that.

We have to face the fact that there is always going to be the "missing technology" problem for astronomical SETI. If SETI astronomers moved up a gear and developed receivers for twisted microwaves and light, any signals they received couldn't be confused with natural radiation because there is no known source in Nature of signals in the form of a multi-twisted helix. But it would be pointless and that's not going to happen. However, this subject is a timely lesson for SETI research: you can't base your research on speculations about what technologies the aliens might be using. If there is more advanced science to be discovered in this universe, as most scientists believe, it will be applied to create more advanced technologies. What we currently see as technologically possible is probably only the beginning of what can be done. Even on planets where high intelligence has evolved and endured, we have to ask: "Where is the motive to send messages that may never be received?" By the time our civilization is ready to mount a continuous program of broadcasts to the stars, one that would have to last hundreds or thousands of years to have a chance of being received (quite apart from getting a reply), we will have our interplanetary probes ready to launch toward the most interesting of neighboring planetary systems. This seems the most reasonable scenario—and it makes searching for evidence of visits during the past few billion years a reasonable strategy.

ANSWERS ON THE MOON AND MARS?

Looking for ET in the solar system is a risky business for a scientist's reputation. It's not quite as bad as chasing UFOs, though it's difficult to see why. If aliens are in the solar system, the main attraction would be planet Earth, but they won't come here and not visit the Moon and Mars. Therefore, scientists at the Society for Planetary SETI Research are scanning close-up photographs of both Mars and the Moon for anything suspiciously artificial, and they have found a few ambiguous structures, though they know these may look quite natural when future missions provide better pictures. But by doing the research now, any interesting sites can be identified for future missions. It doesn't matter if these scientists are wrong because it matters so much more if they are right. In "local SETI" we're not trying to find ET for an exchange of views about the state of the universe. We're trying to find evidence that ET has already found us—a more likely scenario and quite enough to confirm the Grand Hypothesis. But whether we have a chance of doing so or not depends on two things: interstellar space has to be crossable—though probably not by flesh-and-blood astronauts—and evolution on numerous planets has to produce species capable of establishing highly successful civilizations. With so many planetary systems out there, a new civilization might evolve every few million years, though we don't have to rely on a large number of planetary biospheres doing that. Migration from just one advanced civilization over millions of years might take its citizens to many planetary systems.

So what's being done to check on this? Perhaps the most publicized example of what's become known as "planetary SETI" is the work on two NASA photographs taken by Viking Orbiter 1 in 1976. They appeared to show a monumental human face. NASA said it was just a trick of the light that made a very large rock look like a face. But that didn't deter several scientists who worked on the fuzzy photographs for years, ever hopeful that a great discovery could be made. They were encouraged in this by the presence of several structures that were unusual and appeared to be associated with the "Face." Many people thought this was all fanciful and sensational, but what the researchers were doing was developing a genuine scientific hypothesis, knowing that it could be tested by NASA's next missions to Mars. The old "gut reaction" was that better photographs would show a natural rock formation. And, as it turned out, that "gut reaction" appeared to be right in 1997 when the Martian Orbiter photographed the site. But the new photographs didn't settle the issue. The "Face" had given way to a rather unusual rock formation, and the pyramids in the original photographs were still pyramids. So have visitors been building pyramids on Mars? It sounds like a question for the next sci-fi blockbuster, but only better evidence can satisfy everyone that the mystery is solved. Dr. Mark Carlotto, who worked on the photographs for years, said that he would be ready to accept NASA's verdict on the "Face" if this was the only puzzling object. But he pointed out that the "Face" is not on its own. "Nearby are other strange looking objects. Some quite geometrical in shape. A number of them look like pyramids, one apparently five-sided. Moreover, the objects seem to be arranged on the Martian surface in an organized pattern."[13]

Mars also has other suspicious features on its surface that scientists at the Society for Planetary SETI Research display on their website. But as with the "Face," we may be seeing things that aren't there. The human brain is quick to see patterns and can see faces in all sorts of things. Neurological research has revealed a highly evolved face-recognition program running in our brains. Yet a face looking at us from the Martian surface is such an ingenious way to attract our attention that it deserves to be real. If aliens landed on Mars or the Moon after *Homo sapiens* evolved, they should have done something like that. But what I found inconsistent in reading material about the "Face" was that leading researchers were saying it was many millions of years old, judging from the associated geology. This rather undermined their case because nothing like the face of *Homo sapiens* put in an appearance until about two million years ago.

The point I would stress is that whatever the final verdict on the "Face," there is no reason to mock a testable hypothesis. Testable hypotheses in SETI are too rare to be mocked, even if they belong to the so-called fringe. If the "Face" is just a pile of rocks, so be it. Failure is part of the game in science. Most conventional scientific hypotheses fail. But it's not unreasonable to assume that if aliens ever reached the solar system that they might leave evidence on Mars and the Moon. Again the old gut reaction may say "Nonsense," but no one knows—and strange things have happened on the Moon, as some scientists have reported.

THE LUNAR LIGHTS

An alien presence on Mars or the Moon would be consistent with some aspects of the UFO phenomena, though no more than that. The evidence might be millions of years old with no connection to currently reported UFOs. However, if some UFOs are for real their activities are unlikely to be limited to the earth. Consequently, scientists in Ukraine, Russia, and Byelarus have studied lights on the Moon. Scientists at the Research Institute on Anomalous Phenomena (RIAP) in Ukraine work on the provisional assumption that at least some reported UFOs have a physical reality. They therefore investigate any evidence that might support this assumption. "To open a way to the solution of the UFO problem, it should be posed as a normal scientific problem," says the director of RIAP, Dr. Vladimir Rubtsov.[14]

I have corresponded with Vladimir for about twenty years, but during 2007 to 2009 I worked with him, editing his book *The Tunguska Mystery*, which describes a subject he has researched in Russia and Ukraine for about thirty-five years. On June 30, 1908, something flew over an uninhabited region of Siberia and exploded with a force greater than anything in recorded history. Initially it was thought to be a meteorite or comet, but neither explanation is in line with all the data collected by expeditions and hundreds of investigations since the 1920s, much of which is recorded only in Russian. So the study of this mystery continues, especially at RIAP.

Another of RIAP's programs, called SAAM (Search for Alien Artifacts on the Moon), began in 1992 to look for artificial structures on the lunar surface.

The so-called lunar lights were a good start for this research, as they provide a testable hypothesis and are well described in astronomical catalogs. This is because many amateur astronomers have been observing the Moon for more than a century, it being a near and easy world to study with small telescopes. There are good astronomical catalogs of lunar lights because when things happen there—on the face, which is permanently turned toward us—there is a good chance that someone will observe and record them. However, only in recent years has anyone in astronomy had the nerve to associate lunar lights with the UFO phenomena. Most of these "transient light events," as astronomers call them, have been explained away as meteor impacts, volcanic activity, electrical discharges, escaping gas and dust from moonquakes, which of course most of them probably are. But some of the reported lights behave like UFOs, if the reports are correct.

The most interesting lights reported are those that have accelerated rapidly and have made sudden acute changes in direction when traveling at speed. Another puzzle is stationary lights that maintain the same luminosity for up to two hours. "A gas cloud cannot exist as a point object during periods of 10 minutes to 2 hours," says Alexey Arkhipov, one of the original researchers. "And it is unlikely that electrical discharges in a dust-gas cloud can give luminescence of constant magnitude during such a long period."[15] Arkhipov works at the Institute of Radio Astronomy in Kharkov and has in recent years carried out searches, using some of the best lunar photography, for evidence of artificial structures. But on the subject of lunar lights, he has pointed out the strange effects our lunar landing craft seem to have. For example, only one light was seen in the Mare Tranquillitatis before the impact of Ranger 6 in 1964 in the northern part of this region. After the impact it was reported to be one of the best sites on the Moon for transient light phenomena until September 1969. Were the aliens coming to see what we'd just dumped on the Moon? Conventional wisdom would say baloney, it was probably geological activity caused by the crash. But the same sort of thing is reported to have happened between the Sabine and Maskelyne craters after Ranger 8 crashed, and also following the landings of Surveyor 5 and Apollo 11. So if the observations are correct, it's a puzzle.

According to Arkhipov, about 14 percent of reported lunar lights move along bent or broken UFO-like trajectories, with velocities of around one hundred kilometers per second and accelerations similar to those of UFOs observed near Earth." He explains that some UFO events on Earth, given that reports are accurate, would be observable from the Moon through conventional telescopes. Another claim is that the presence of lunar lights seems to have coincided in time and abundance with the waves of reported UFO sightings on Earth, especially the UFO waves of 1954 and 1967. Of course, no one can conclude from this data that aliens are active on the Moon. The reliability of the data can't be assessed. However, this subject is wide open to testing. Perhaps we could see the UFOs at our future landings and crashes on the Moon, if the reports of lunar lights in the past are true. After all, we might expect any "intelligence" behind the "lights" to observe our spacecraft that hit the lunar surface—or land on it. We may or may not observe lights clustering around new landing sites, but it will cost next to nothing to look. The spacecraft would be traveling to the Moon anyway.

All this seems bizarre: UFOs checking our lunar landings. But that doesn't matter, always providing the idea can be tested. No one is saying: "There are these strange lights on the Moon that are alien spacecraft." What is being said is: "These lights, according to many astronomical reports, are appearing on the Moon and there is no clear explanation for them." They may arise from perfectly natural phenomena, but let's not waste a free chance when the next vehicle lands on the Moon. Let's test the hypothesis that there's something odd to investigate. Moderately large optical telescopes could do the checking, plus the many keen amateur astronomers who could take a look and make some records. And if lunar lights do consistently appear around new landing sites, we would have a problem to solve. So what would we do?

This all looks like another scenario for a sci-fi movie, explaining how the UFO mystery was solved on the Moon. But imaginative film directors should keep in mind two things: If alien artifacts are visiting the Earth, they will also be visiting the Moon. Secondly, the Moon offers an unsuitable environment for biological beings but a benign and constant environment for robots. There are no interfering life-forms, no difficult weather conditions, just a light level of radiation most of the time with the odd burst of intense radiation during solar flares. Nothing to worry a top-class robot. And the splendid view of the entire Earth, as it rotates on its axis, should make up for any lunar problems. So the Moon could have had visitors during the past 350 million years, since the ozone layer of the Earth's life-supporting atmosphere was well established and detectable. And well-constructed buildings on the Moon would last a very long time. Nothing much has happened there since the great bombardment of some three billion years ago, except for the occasional meteorite. The Moon is heavily scarred by impacts, but most of the big stuff hit it in the early days of the solar system. Major impacts have been well spaced out since then. You might spend a lifetime on the Moon with no fear of being hit by a meteorite. So I reckon the Moon is the best bet for alien artifacts, especially since evidence could be as old as the oldest fossils of land life on Earth, an immense time frame during which alien missions might have arrived.

In recent years Alexey Arkhipov has gone on from the study of lunar lights to the search for artificial structures on the Moon. This looks to be a better bet than checking on lights that have an inconvenient habit of rapidly disappearing. The remains of lunar bases, on the other hand, if there are any, would not easily disappear. Arkhipov has used photography from the lunar orbiter Clementine. "About 80,000 Clementine lunar orbital images have been processed, and a number of quasi-rectangular patterns found," he says.[16] The search was for unnatural patterns, mainly straight lines and rectangles. Obviously, a vast number of digital photographs can't be scanned manually for signs of artifacts, so Arkhipov and his colleagues had to prepare a computer program for the initial search. Arkhipov tells us that after all the sorting out, the computer system ignored 97 percent of the images, and after geological considerations only eighteen photographs remained interesting candidates. As Arkhipov says: "Of course, some or all of our finds could be geological formations. But the possibility that they could be archaeological features is so important that it should not be ignored *a priori*. Ultimately, only human exploration of the Moon will determine whether these features are artificial or natural in origin."[17] This may not be entirely correct, as in 2011 NASA put its Luna Reconnaissance Orbiter into an elliptical orbit, which took the craft to within thirteen miles of the Moon's surface to photograph the Apollo landing sites. The evidence of the landings is clear to see in numerous photographs on NASA's excellent website.

Arkhipov has had the same problem with the science establishment as those scientists working on the UFO phenomena: "Although the *SETI League, Society for Planetary SETI Research*, and the *Russian SETI Center* support these studies, few scientists dare to search for evidence of extraterrestrial intelligence on the Moon."[18] The problem is that everyone not familiar with the science that supports astronomical SETI thinks this area of research is beyond the fringe. But when you consider the relevant mainstream science, which I hope this book provides, it all looks reasonable. It just rests on evolutionary developments on other worlds and the possibility of interstellar transport. In our case the evolutionary pressures provided by dangerous open country instead of relatively safe forests came at the right time to make *Homo sapiens* from one species of upright apes. But those upright apes had to be present to be subjected to the evolutionary pressures of the time. Similar situations may be rare in the histories of life on other worlds. But if the evolution of technological intelligence is frequent enough and interstellar travel possible, then intelligent entities or their artifacts (including robotic astronauts) may have left evidence on the Moon and Mars. It's another long shot, of course, but the data from science tells us that there's been more than 350 million years for potential visitors to find the life-supporting presence of the Earth.

ARTIFACTS ON EARTH?

Although erosion soon hides the past on Earth, scientists at RIAP have been interested in what they call "paleo-visitology": when something that just might be an alien artifact turns up, it becomes the subject of scientific scrutiny. A few mysterious items have been found, publicized, studied—and are found to have nothing to do with visiting extraterrestrials. But very strange objects have to be checked out just in case. The difficulty with this line of research is that dry land on Earth is subject to erosion and is constantly changing, while the period is immense during which visitors may have left bits and pieces behind. You can see the problem by looking at paleontology. The fossil of a previously unknown herbivorous dinosaur is discovered that a hundred million years ago must have roamed our planet in vast herds for several million years. Yet all the paleontologists find are a few bones from a few individuals that don't even make up a complete dinosaur. It's the same with most species that once flourished in great abundance on our planet. There are great gaps in the fossil record that one day may be filled. But accidentally leaving things behind when you say good-bye to an alien planet is very different from marking the occasion of your visit. My guess is that visitors who wanted to leave a calling card would have left it on the Moon, and several people right back to at least the 1960s have guessed the same. Mars would be less suitable, having some erosion and being a lot farther away, but it seems a more interesting world to visit.

As for contact with intelligent aliens in historical times, say, during the past few thousand years, there are historical records and ancient mythologies that tell us about mysterious visitors, but none that provide us with information we do not know but for which we could test. A model that looks like an aircraft found among ancient ceramic art tells us nothing. There are many such claims to past visits: carvings and drawings interpreted as rockets and spaceships and people in astronaut gear. One thing you can rely on: if something is interpreted in terms of our current technologies, then it does not have an extraterrestrial origin. Any alien artifact on Earth will have crossed interstellar space and should be beyond anything familiar to us.

LANDINGS ON ICE

Winter in RIAP's part of the planet, Ukraine, has provided some special evidence. It was January 7, 1990, at about 8.30 in the morning, when a Mr. Vorontsov, an inhabitant of Merefa, a small town thirty kilometers from Kharkov, saw a large "top-shaped object" about twenty-five meters in diameter. It was on the ice of a nearby bay or hovering just above it, and its base was pulsating. After about ten minutes it flew away. The RIAP report says: "A big round piece of ice on the landing site sank into the water, then rose again to the surface." A pattern of concentric rings a meter wide were marked in the ice, the outer ring being almost twenty-two meters in diameter. Two days later, Vladimir Rubtsov and Pyotr Kutnyuk arrived from RIAP.

They took samples of ice from the circle and controls, but chemical analyses showed nothing significant. What was most interesting, however, was that at the time the circle was formed a period of warm weather had made the ice too thin to walk on, making a hoaxed circle an almost suicidal task, as the water beneath was eight meters deep. The RIAP report says: "It was not until January 13 (six days later) that it became thick enough to safely bear the weight of a man walking on the ice."

Another ice circle was reported in the same area in the winter of 1995. A Mr. Mandych, in a village close to the original site, was smoking and looking out of his window on the night of December 4–5 when he saw a "big round object" flying noiselessly toward the area of the 1990 ice circle, an event well known to everyone in the village. Next morning, he and other villagers went to the bay of the Mzha river and saw a circle some twenty-six meters in diameter that had been cut from the ice. "It slowly rotated under the action of the river current," says the RIAP report. "On either side of the circle's edge one could see a few narrow concentric rings scratched on the surface of the ice." As before, it was a dangerous place for a hoaxer to make a circle. "The ice could not have supported even a dog," said one witness. But when RIAP scientists visited the site a few days later, the ice had thickened to about ten centimeters. The circle was frozen into the surrounding ice, although the outer and inner concentric rings were still visible. Chemists analyzed samples, but nothing significant was found.

Ice circles of various sizes have also been reported in Europe and North America, and chemists have analyzed samples, but again, nothing significant has been found. What one would like, of course, is a chemical pattern associated with such circles—a pattern that is repeated and therefore predictable. But whatever or whoever forms these ice circles doesn't leave anything for science to get hold of. And that's what we need: phenomena that give us something to grasp and follow up.

PHOTOGRAPHIC EVIDENCE

At one time hardly anyone had a cine camera to hand when a UFO flew by, making moving images of UFOs very rare. Then camcorders became a popular accessory of modern life and the number of strange craft and puzzling lights on tape increased. More recently, camera and video facility have been added to mobile phones, making it even easier to record strange aerial craft and UFOs. In fact, there's enough visual data available to keep photo analysts busy for the rest of their lives. Of course, this area of ufology has not been short of hoaxers—mostly incompetent hoaxers, although some fakes have fooled the experts.

What has been recorded show strange and spectacular images: luminous spheres, metallic discs in daylight, fast-moving zigzagging lights, cylinders, and other weird objects. There are numerous collections to explore, and many are freely available on the web. Jaime Maussan, the director and anchorman of the Mexican news program *60 Minutes*, has gathered more than five thousand tapes since a large crowd in Mexico City in 1991 came to watch a solar eclipse. They were there to see the eclipse but instead witnessed a UFO that was captured on tape by several people with camcorders and by the television cameras of Televisa. Since then Mexicans have recorded lots of UFOs both in daylight and at night, and have sent them in to Maussan's program. One Mexican priest, seeing the theological implications of visitors from other worlds, was said to spend half his time on the top of his house with a camcorder. Today, all the videos are in a special library in Mexico City. We don't know what proportion could be genuine, but that archive exists for scientific investigation. And it's the same for thousands of other videos. The best material awaits rigorous study by specialists at major universities whose conclusions would carry weight. Of course, the "best" material would have to be carefully selected. The data shows that a well-publicized case induces people to grab their camcorders and mobile phones and search for anything that might be a UFO.

So are the aliens here? And if so, why are they so keen on Mexico? Although the camcorder and mobile phones provide moving images, which is a plus over still photography, these images are recorded electronically and digitally, which means that they can be altered by the same technology. The problem is that there is so much material and so few video experts to eliminate the hoaxes. The work of a video analyst can be highly professional, as readers can see from a report by Jeff Sainio (staff photo-analyst with the Mutual UFO Network) on the video taken in Israel in 1999. Yet the conclusion of one analyst, no matter how well qualified, is never enough. A video taken near Alicante in Spain stood up to several independent analyses until it was shown to be a teddy-bear balloon.

Videos might provide serious data, but the work has to be coordinated in some way so that qualified people at several universities are involved. Videos that initially pass the technical tests to detect fakes could be examined for the dynamics of UFO flight and evidence of any physics and chemistry that might be consistently observable in different videos, such as color contrasts resulting from radiation chemistry in the atmosphere. Of course, there may be nothing to detect, but without serious research we can never know.

Analysts tell us that it can take a hundred hours or more to analyze just one short video sequence, which explains why videos are seldom checked before they appear on television. It's like the "cold fusion" story we considered earlier. Several teams of top-class physicists would have had to confirm the original findings for the scientific establishment, and society in general, to accept the reality of "cold fusion." Its existence as a phenomenon of physics could have solved the world's energy problems, which the science establishment would probably consider rather more important than testing for the universality of life and intelligence by studying the evidence for UFOs being possible alien artifacts.

Before we leave the subject of photographic evidence, Vicente-Juan Ballester Olmos, a leading ufologist in Spain, has established FOTOCAT, a comprehensive catalog of UFOs on film, tape, and chip. It has some five thousand items from almost every nation, so it's a promising development, providing material for scientific research that might produce results that everyone could accept.

MYSTERIOUS LIGHTS

The mysterious lights of Hessdalen in central Norway have provided a scientific puzzle since the mid-1980s and a lot of unjustified speculations. But in more recent years they have attracted scientists with the necessary equipment and experience to get to grips with the various phenomena on display. It should be said that Hessdalen is not unique, although it is the most publicized place where strange luminous phenomena have been regularly observed. What look like plasmas that defy the laws of physics and other luminous objects are reported in some thirty different parts of the world. Some have received scientific attention starting as long ago as 1963 with the French Lights in the Night Society. Perhaps the best-known mystery—apart from the phenomena at Hessdalen—are the recurring lights in Piedmont, Missouri, that were studied by Harley Rutledge, a physics professor at Missouri University. Working with his students, Rutledge's research continued for years, but nothing was proved except that the lights were inexplicable.

In recent times, astrophysicist Dr. Massimo Teodorani of the Institute of Radio Astronomy in Bologna, Italy, has made a comprehensive assault on the problem and has already led five scientific missions to Hessdalen, as well as studying the Earth-light phenomena elsewhere on the planet. "There are," he says, "about 30 areas of the Earth where such anomalies are present in a recurrent way. The Norwegian area of Hessdalen is the most important where visual, magnetometric, radar and radiometric recordings have been carried out since 1984."[19]

In 1984 a research station was set up at Hessdalen, and a group of Norwegian researchers, led by engineer Erling Strand, began to collect the observational data that would eventually attract international attention. Scientists from several countries, including Italy, America, and Britain, have been involved with the research to try to solve this major mystery. In recent years, Strand and his colleagues have established the International Earthlight Alliance, which has a comprehensive website reporting on the latest thinking and research. Strand and his colleagues say the differences in the phenomena are so great that they could be dealing with different things, and research seems to confirm this.

From the many observations, they have divided the phenomena into four categories:

1. "White or blue-white flashing lights." These appear near the tops of mountains and usually last only seconds, but they can flash several times while they are moving.
2. "A yellow light with a red light on top." The yellow light varies in intensity, while the red light can flash on and off. It is thought that the change in color is due to shifts in frequency (Doppler shifts) of light from this phenomenon. But how this happens is a complete mystery.
3. "A yellow or white light," the most often observed light at Hessdalen. The researchers say it can remain stationary "for more than an hour," though it also moves around slowly over the landscape.
4. "A black object with lights" has been reported—and once photographed—near Hessdalen. A similar object has also been filmed in Hungary. This is the "odd item out" as far as Hessdalen is concerned, and not something amenable to scientific investigation. About the plasma lights, Teodorani says: "The luminous phenomenon behaves like a thermal plasma with a temperature around 6500 degrees Kelvin."[20] This is about as hot as the surface of the Sun, and research indicates that this temperature remains more or less constant. Increases in luminosity—which might be due to increases in temperature—seem to result from the unexpected way these plasmoids increase in size. "Luminosity increases in a drastic way because of the sudden apparition of many smaller light balls around a larger luminous core," says Teodorani. "Some of the secondary light balls are often ejected from the core."[21]

This obviously cries out for an explanation. But before this strange structure was revealed, several mechanisms of physics were suggested for the Hessdalen plasmas, including mini black holes, anti-matter, ionized gas confined by a rotating magnetic field, and seismic phenomena. And in 2003, physical chemist David Turner came up with a theory for ball lightning that involves the generation of electricity by geophysical forces. This has helped to explain at least some of the Hessdalen phenomena. The reality of ball lightning (usually football-size plasmas that drift along until they fizzle out or explode) has long been accepted, although physicists had no explanation. Some people have been burned, and astronomy professor Roger Jennison, when a passenger on an airliner, once witnessed a sphere of ball lightning drift down the aircraft's gangway. Sometime after the event, I spoke with him about his unique experience, and he was then still trying to work out the physics—without success.

"This longstanding problem," says Turner, "is completely eliminated once it is accepted that a plasma is both a phenomenon of physics and a mixture of chemicals."[22] In other words, trace elements in the atmosphere plus high temperatures make persistent plasmas possible. "Turner's thermo-chemical model," says Teodorani, "is able to explain as a natural phenomenon at least 80 per cent of the data collected by us."[23] However, there are some phenomena that do not show the spectral signature of a plasma. "In such cases," says Teodorani, "a clear signature typical of uniformly illuminated solids has been found."[24] This obviously presents an even bigger unresolved problem.

For a couple of decades an "extraterrestrial probe hypothesis" was offered for a small proportion of the Hessdalen lights, but hypotheses lead to progress only if they are confirmed, and, as far as the science community is concerned, tests have not confirmed anything extraterrestrial. However, strange and inexplicable phenomena justify scientific attention anyway, and the first task must be to try to explain all phenomena in terms of current science. It's the only rational route into the unknown. We therefore have to wait to see if some of the phenomena at Hessdalen and similar phenomena elsewhere remain inexplicable in terms of known physics and chemistry. If so, those in astronomical SETI might find that the evidence they seek is here rather than among the stars. In an e-mail to me, Dr. Teodorani said: "If we humans are already able to produce nanotechnology and to plan DNA super computers for the very near future, I wonder if a highly evolved civilization is able to instruct elementary particles (such as electrons and ions) in the same way in which we are currently able to instruct a silicon chip. In such a case an apparently intelligent plasma ball could be a sort of robotic probe itself."

This idea is certainly in line with the situation on Earth, where the most advanced nations possess the most advanced "sensing technology." The information-gathering devices of visiting aliens could therefore be more advanced that anything we could envisage, so we would probably not recognize them as such. Meanwhile, quite apart from the extraterrestrial factor, a more attractive context for research could hardly be imagined than those plasmas with startlingly uncharacteristic plasma properties. So the task is to try to explain all the phenomena in terms of known science and to duplicate the various plasma phenomena in suitably equipped laboratories. There may not be intelligence behind those mysterious lights, but there certainly will be new physics.

ABDUCTIONS AND ARTIFACTS

People have actually insured themselves against being abducted by aliens, but have they thought about collecting the insurance money? The aliens might decide to keep them. Or, far more likely, the insurance companies would be awkward, as usual, and ask for proof. And since no one anywhere has yet been able to prove an alien abduction, this new venture looks like a winner for the insurance companies. However, although this appears to be a no-win deal for anyone worried about alien kidnappers, anyone who could actually prove an alien abduction would be far more than a real winner, since they would have confirmed the major hypothesis in science. So let's consider the possibility that there are alien abductions to confirm, even if we don't believe that there are, which through lack of physical evidence seems the rational position to take.

Of course we might expect visiting aliens to collect samples of life on Earth, but why abduct large numbers of *Homo sapiens* at this particular time in Earth history when they could have captured plenty of *Homo sapiens* during the past 250 thousand years without making headlines in the newspapers? And why are there so many similar physical investigations in flying saucers? Most abductees are asleep at night when abducted, and all wake up safely in their beds in the morning ready for the coming day. You couldn't make up a less likely scenario. However, some people are abducted in daylight from their cars or when out walking, but most of these experience what's been called "missing time," so it's not until much later—after they've been hypnotized—that they find out that the aliens have been mistreating them. With so many people being abducted (an estimated two million in the United States alone), you might expect abundant forensic evidence with no shortage of saucer dust in clothing. The abduction ufologists do claim forensic evidence, such as particles, hairs, stains, jelly-like substances, and implants, but this "evidence" is just reported and re-reported in the printed media and on television. There are no reports from forensic scientists to support belief in any alien abduction. There have been tests on implants surgically removed from the bodies of abductees, but the laboratories have been understandably ambiguous. Some abduction researchers have voiced disappointment that analysts have not drawn conclusions from testing abduction implants. But what else can they expect? In one case the Los Alamos Laboratory stated "origin unknown" nineteen times in a long report. The problem is that to make progress you would need to discover a pattern of evidence from testing many suspected implants—or any other alien artifacts. "One-off" findings can only be a start to systematic and continuous research. What someone needs to find is an interesting pattern of physical evidence, if such a pattern exists to be found. That would be the time to take the abduction phenomenon seriously.

But this approach doesn't fit well with the media. In television programs interviewers ask witnesses about their experiences. Actual alien artifacts are rarely available to show viewers. Yet even when physical evidence is available it seldom gets the required scientific treatment. By contrast, the first rock samples from Mars will be analyzed and studied by dozens of scientific institutions. Yet Mars samples, if they contain microbes, fossilized or alive, would at best confirm the universality of life, whereas an alien artifact, say, something pinched by an abductee from a flying saucer, could confirm the universality of both life and intelligence. The immensity of such a claim would therefore demand reports from the most eminent scientific institutions. That's what journalists should expect from UFO stories that offer claims of contact with aliens and their artifacts—and that's what they never get.

Not long ago I watched a television presenter on this path to ambiguity. She had discovered Bob White of Reed Springs, Missouri, who in 1985 had witnessed an object fired from a UFO. He recovered the object, a sort of missile about eighteen inches long, and he offered it for tests to be shown on television. Actually it had already been tested previously at five different laboratories and had been found to be 80 percent aluminum, but the television reporter didn't tell us that, or that analysts doubted its authenticity as an alien artifact because its composition was similar to a commercially available alloy.

Analysts thought it might be a piece of space debris that partially melted during its descent from orbit, though how it came to have its strange structure was not explained.

However, the television reporter started to make the program interesting when she disclosed that Bob White's missile might not, after all, be the usual "one-off" find. In 1947 a similar object had been recovered in Denmark by the US Counter Intelligence Corps, and a detailed drawing of this object had been released to the public in 1998 under the Freedom of Information Act, though the object itself had apparently been lost. No one can tell you where it is, but the drawings, which are on the web, do show something like Bob White's missile. Obviously, it would be more than useful to have this second object analyzed. If its chemical composition matched Bob White's missile, and it had the same ratios of isotopes, we might be reaching a eureka point. Unfortunately I was discouraged about this possibility on reading a press release from Bob White's supporting group. A scientist, someone said to be close to officials charged with keeping the reality of UFOs secret, claimed to have seen the second alien missile in the course of his work for the government. The possible reliability of this information was then completely undermined when the scientist was quoted as saying: "We have reversed engineered alien technology for our benefit." Credibility flew out the window and vanished completely when we were then told that the Germans were working on flying saucer technology during the Second World War, but that Hitler didn't trust the aliens—or was it that the aliens didn't trust Hitler? What more can one say about this source of information? Only that a good education in science does not necessarily ensure that a nut won't come loose.

In the same television program, Dr. Roger Leir, a surgeon in the United States, was claiming a possible extraterrestrial origin for eleven implants he had removed surgically from abductees. The presenter should have realized that if Bob White's missile was fired by angry aliens and Roger Leir's implants were extraterrestrial, then the major hypothesis in science had been confirmed and someone had better get ready to visit Stockholm and collect a Nobel Prize. Also, the need to spend billions of dollars in the coming decades for NASA scientists and others to test the Grand Hypothesis would no longer exist. The hypothesis would have been tested and confirmed by low-cost chemical analyses. That show's presenter faced the biggest scoop in the history of journalism, but only if Leir's implants or Bob White's missile could be proved to be from out of this world.

Dr. Leir has given many interviews with the media and has published a couple of books, though research papers to enable scientists to assess his work are needed. Nevertheless, he has sent the surgically removed implants to several labs for analysis because he genuinely thinks it possible that they have an alien origin. One lab reported that the implants are composed of substances comparable to those found in meteorites, a finding that was bizarre since they had come from human bodies—and also meaningless since meteorites vary considerably in their composition. The other reports were not much help either. It was the same old story: nothing significant was found. The problem is that unless an artifact is obviously alien, like a crashed saucer or a biological alien, it's going to be a search for chemical evidence, which would need a lot of work.

What often happens is that ufologists send suspicious objects to labs that report that the objects are chemically in an unexpected pure state (like the fragments of magnesium that were collected after a saucer exploded over a beach in Brazil). Sometimes they are in an unusual state, but without a pattern of chemical evidence from a number of different cases, that's not really interesting. The worst reports, which ufologists sometimes offer as evidence, are those that state that samples are composed of substances unknown on Earth. What a lab really means by this is "substances unknown to the lab." There are millions of different substances on Earth, and no one can know them all. All this tells us is that little can be concluded from one-off findings. However, there is a scientifically legitimate way to tackle the problem. Nature offers a basis for testing any possible alien implants or artifacts. The testing would be based on the proportions (ratios) of stable isotopes in the artifacts. This approach depends on stable isotopes, not radioactive isotopes. Many elements exist in slightly different forms although they are the same chemical element. The differences lie in the number of neutrons they have, which makes for differences in mass. Take the element magnesium. It exists in three forms, having 24, 25, and 26 units of mass. And the proportions in which it exists on Earth are 79 percent, 10 percent, and 11 percent. Carbon has two stable isotopes and one radioactive isotope. Magnesium and carbon are, of course, just two elements of many that have isotopes that might enter manufacturing processes on another world. Consequently, alien artifacts should carry the isotopic signatures of the places where they were made. In our case everything was in a molten state after the Earth formed, and there was lots of thermal activity for a long time. Consequently, the chemical elements were well mixed, which has produced some consistency in the ratios of chemical isotopes. But the ratios of isotopes would be different for other planets in other planetary systems. First, the orbiting discs from which planetary systems form would not be chemically identical. Second, once planets form, the thermal and geological forces on chemical elements would vary. Chemistry therefore offers a way of testing the credentials of alien implants that have been offered as evidence of alien abductions, or of any other alien artifacts on offer. The beauty of this chemistry is that although unstable radioactive isotopes change over time, the ratios of stable isotopes will remain more or less constant. Manufactured items from Planet X—where the alien visitors live—should therefore carry a consistent chemical signature of that world. The chemists checking possible alien artifacts would use the technique of "mass spectrometry," which can determine isotopic ratios with high precision. So, have any supporters of alien abductions got around to this line of scientific investigation to prove their claims? Well, I think a few may have thought about it, but there are no reports available as far as I know. And, more importantly, there are no research papers of the sort scientists expect to be available: papers that describe precisely how research was carried out and the results obtained. Researchers on the fringe used to say that the specialist journals would not publish their papers. That was true. An editor with the journal *Nature* once told me that you needed a university address just to get your work considered for publication. But with the development of the web such barriers no longer exist. The old and trusted system of peer review won't stop your papers being read. Yet if scientists are to take them seriously the same standards of presentation apply. Those individuals who believe they've been abducted may be sincere and perfectly normal people, but without proof the science community will continue to think they are deluded, especially as psychology offers an explanation. "Sleep paralysis" is a widely studied phenomenon throughout the world—and a common phenomenon. Different countries have different traditional visitors during sleep paralysis. Demons and witches may invade dreams in the East and carry the paralyzed sleeper away. But these days dreamers in the United States and Western Europe get carried away by aliens. All this tends to make the alien abduction phenomenon the stuff of modern folklore, something that has developed since people began to report discs in the sky in the late 1940s. But those strange structured craft reported by many aircrews provide an entirely different category of data from what abductees describe from a psychiatrist's couch.

Mainstream society doesn't take alien abductions seriously. When people are kidnapped by *Homo sapiens* rather than aliens, the police are called in, who then bring in the forensic scientists. These are the people with the tools and know-how to find biological, microscopic, and chemical evidence. But what happens if you're abducted by aliens and call in the police? The police are more likely to abduct you than search for law-breaking aliens. Alternatively, it's the hypnotist's couch—not the police—from which you'll tell your story. Many abductees of course don't remember they've been abducted until months later, which is a bit late for the police to catch the aliens, though the clothes being worn at abduction time should still provide evidence. But no one seems to have done a full forensic study of the clothes worn during abductions—usually night clothes. So there are plenty of stories but no forensic chemistry. Another odd thing about abductions is that the aliens seem to prefer American citizens. No other nationality holds such attraction for them. Dr. Vladimir Rubtsov tells us: "As for the Soviet Union, saucer abduction reports from its territory were extremely rare." According to reports, people in that part of the planet just had friendly chats with the aliens, who never forced them into their craft for medical experiments—and never said anything interesting either, although the aliens spoke passable Russian.

So it's up to the abductees—or rather their supporters who write books about them—to scientifically confirm an unlikely phenomenon. Future abductees who want to mark their place in history should ask the aliens to tell them something new, something we could test for. This is not asking much for being an experimental guinea pig in a flying saucer. Actually, with the number of abduction claims, someone by now should have pinched an obvious piece of alien technology. Either security on board flying saucers is totally tight, or abductees are totally honest—or totally deluded. The hard fact is that you can't steal anything from dreamland or report anything from a dream that is not already in your own head. Abductees provide information but nothing too demanding for an average Earthbound imagination. We need something that would knock us flat if we found it to be true. We need the sort of information we could give our ancestors if we traveled back in time. And we wouldn't have to travel back far to astonish everyone—say, a couple of hundred years. We wouldn't provide the sort of information abductees and contactees offer. We wouldn't tell them how to save the planet and reform society. To convince our ancestors that we came from the future we would provide insights into the nature and use of electricity and magnetism, or some equally interesting phenomena. This would be information that our ancestors didn't possess but that they could test for—and probably use. Yet what abductees offer is all based on what we already know. The scientific position is that hypnosis is not a tool for obtaining reliable information. It doesn't really demonstrate anything except that the people being hypnotized all seem familiar with the same mythology. Most professional psychologists would say that the abduction phenomenon is psychosocial, yet major UFO conferences continue to feature lectures by abductees and by the hypnotists who study them. Quite apart from hypothetical implants, forensic studies of their clothing when abducted could be far more interesting than their stories.

I know that one leading psychiatrist, the late John Mack, a professor at Harvard University no less, came to believe that alien abductions are taking place, but like everyone else involved with this subject he had no information from his sessions of hypnosis that scientists could check, and certainly no significant forensic evidence. However, he may have been misled—as many people have been—by claims of physical evidence. He has said that there is "highly robust physical evidence that accompanies some of the abductions." So where are the scientific papers that indisputably link this evidence to alien activity? Mack has also made the point that abductions also take place in the daytime when people are not asleep. But in such cases abductees suffer from the so-called missing time, during which their memories are a blank, and, like the people abducted while asleep, they don't recall much until weeks or months later when they reach the welcoming arms of the abduction specialists and a session with a hypnotist.

The bottom line of this distracting phenomenon is that if you accept any alien abduction story as true, you also accept that the major hypothesis in the history of science has been confirmed. Such an advance in established knowledge cannot be made without totally irrefutable evidence. And if the abduction specialists want to begin to interest the science community they need plenty of physical evidence or information that we don't know but for which we could check. However, one leading abduction specialist, Derrel Sims, dodges this requirement. He has said: "If the devices are alien in origin, we will find no discernable [sic]technology in the embedded devices (implants). Nothing has been found to date."[25] So what could the labs look for? They could only look for isotope ratios that are not consistent with the implants having their origin on Earth.

It's therefore understandable that the lack of real evidence has led most people to conclude that the phenomenon of alien abductions is part of a mythology that has developed from more credible UFO reports. But is there a minute kernel of reality lurking there? I don't think so, but I approve of any scientific work that the abduction specialists carry out to try to find out. Yet our history since the earliest written records shows that human beings are highly susceptible to unfounded beliefs, and we don't want more to pervade our world. Since *Homo sapiens* established science, it has stood against unfounded claims and shown by its success how knowledge can be built up for its own sake and for the development of new technologies. Anyone is welcome to play in the game of science, but you have to play according to its tried and tested rules. The claim that people are being abducted in large numbers by aliens is so extraordinary, and has such colossal implications, that it would need the most impressive evidence for it to be taken seriously. And claims without proof are going to undermine scientific interest in the more credible aspects of the UFO phenomena, where any consideration of the extraterrestrial hypothesis should be based on SETI science and not on a growing mythology. There's a vast difference between information that comes from individuals under hypnosis and report from pilots and aircrews who are not usually in a state of hypnosis—reports that are sometimes corroborated by onboard and ground-based radar and photography.

BRING IN THE FBI

"The Skeptics UFO Newsletter" tried to encourage abductees to prove their stories by offering $12,000 to any abductee who could do so. No one tried. The newsletter pointed out that kidnapping is a federal crime in the United States and the FBI is bound to investigate any claim of abduction it receives. Yet no abductee has lodged a claim with the FBI, as far as the "Skeptics" know. Perhaps this is because the FBI would not rely on evidence obtained by hypnosis. Abductees would be told that if the FBI investigates and finds a claim false, the claimant has to pay a fine of $10,000 and may even go to jail. So there's an obvious risk involved.

A spokesman for the "Skeptics" told me they would want to do the investigating themselves before handing over their $12,000. However, any abductees with proof should challenge the "Skeptics UFO Newsletter" rather than going to the FBI. At least they won't be fined or sent to prison, though, according to their spokesman, the staff at the newsletter might want a ride in a flying saucer before handing over the money.

Abductions seem to be a historical extension of the contactee phenomenon that was all the rage in the early days of ufology. Meeting an alien for an exchange of views was a quick way to become a celebrity. But most contactee reports were riddled with naïveté. The contactees didn't know enough science to make their stories credible. They said the aliens came from the other side of our galaxy or from another galaxy, which is a long way to have them travel when there are thousands of sunlike stars nearby. And the aliens were always like us. One old rascal who claimed to be a contactee offered the media photographs of men from Mars and Venus and actually got them published—I saw them. I also saw this individual on television describe his trip to the Moon before the Apollo Program. He claimed to have seen a lunar dog on the other side of the Moon, a remark that shocked astronomer Patrick Moore, who was interviewing him at the time. He was not alone in his fantasy. Before NASA had discovered the true nature of Venus and Mars, the media had promoted several people who used to talk with the inhabitants of those worlds. But it became rather difficult to do this when Mars had practically no atmospheric oxygen and the surface of Venus was hotter than molten lead. So, if you wanted to talk with the aliens, they had to come from farther away, as all those different humanoids featured on the web seem to have done.

There's just one more development in our subject that weighs heavily against the credibility of contactee and abductee stories. Samples from Mars that may contain alien bugs are soon due to land on Earth. Obviously NASA doesn't want a plague of Martian flu to follow their mission, so there will be a highly complex facility ready to contain the samples until they are found to be microbe free. It will be a tight operation. No alien bugs will get out and no Earthly bugs will get in. NASA says that just the planning of the facility took seven years and cost millions of dollars, so the abductee specialists who publish best-selling books on their subject could save NASA and American taxpayers a lot of money. If alien abductions have taken place, the bugs are already here in quantity, yet no abductee has yet succumbed or passed on an alien infection. So could they convince NASA that it's going to a lot of unnecessary expense—because so many contacts with alien biology have already taken place? I guess not. NASA would want convincing evidence, and there isn't any.

GOING AROUND IN CIRCLES

Many people have spent pleasant summer days scrutinizing circles in the cornfields of Southern England, where large and complex patterns have appeared, often overnight, without any sign of hoaxers at work. No one has been caught in the act of circle production, which is more than strange after so many years of circles manifesting themselves in well-populated regions, such as those in Southern England. All-night summer vigils in the fields by enthusiastic investigators, waiting for crop circles to form, has gone nowhere, though plenty of articles exist on the subject, as do many photographs of beautiful patterned circles taken from the air. A large number of circles are so astonishingly beautiful when viewed from the air that whoever makes them must have artistic genius or contact with such a source.

But although scientists have admired these artistic creations, very few have entered the circles to carry out research. The amount of published scientific research on crop circles is minute, which is surprising when this subject has been a scientifically accessible mystery for decades. Like all subjects with a connection to the UFO phenomena, the crop circles have attracted crowds of enthusiasts yet few investigators with scientific know-how and laboratory facilities to really investigate the phenomenon. What has kept the scientists away is that many people have believed that aliens are responsible for the best crop circles. These believers ask if it's credible that an army of highly artistic circle makers create the most amazing patterns in the cornfields year after year and never get caught in the act. Of course, with a mystery like this there's no shortage of nutty ideas. Some enthusiasts claim that crop circles are the work of Earth spirits and universal psychic forces, so following the same path as many UFO investigators. They try to account for one unknown (crop circles) by another unknown (the paranormal and psychic forces). One person actually wrote: "They're due to psychic projections from human beings—dead or alive. Past minds are being channelled into cornfields." So it's not surprising that most scientists wouldn't be caught in a crop circle—"dead or alive." That's a pity because a few bold scientists have found enough physical and chemical evidence to justify mainstream scientific attention.

The Center for UFO Studies (CUFOS), the leading organization for the scientific study of the UFO phenomena, has stated: "There is still no concrete evidence that crop circles are related to UFO sightings." Nevertheless, the belief that there is a connection is still strong. So, are superior aliens playing games with us? In science the unexpected is always just around the corner, but to suggest that alien visitors are testing our IQs with complex patterns in cornfields is to move far beyond the fringe. Actually, though, it's not completely crazy: patterns and signs have been the basic means of communication throughout human history, so diagrams in the crops would seem to be a creative way to communicate. However, the situation is one for science. Hoaxers are definitely known to have produced messages in the crops, so the focus has to be on the chemistry of the soil and the cell biology of the affected plants. Hoaxers couldn't produce such evidence.

Although crop circles have hardly been touched by the tools of science, some scientists have banded together to carry out research and publish in the scientific literature. And their research seems to show that "genuine" circles have physical characteristics not found in hoaxed circles. The leading crop circle research group, the BLT Research Team, states that about 90 percent of the crop circles the it has studied show similar plant and soil abnormalities, while known manually created circles do not. ("BLT," by the way, are the initials of the founding members of the "Research Team"). You might therefore think that university biology departments would move into the subject, especially as crop circles are wide open for scientific study, staying put as they do while you take samples and make measurements, unlike flying saucers, which are rather difficult to pin down. But no. Crop circles are too close to UFOs for academic comfort and are best avoided.

The main biological evidence involves the "growth nodes" on the stems of cereal crops. The nodes are like tiny "knuckles" at places along the stems. They give plants support and enable them to bow toward the light. And in circles thought to be "genuine," researchers have found them to be elongated and sometimes ruptured. So what's the mechanism responsible for this effect? Dr. Eltjo Haselhoff, who has published a paper on his research, makes the case for microwave radiation: "Although there are known biological effects that can create node lengthening, these could be easily ruled out," he says. "It was clear that something else had happened. The effect could be simulated by placing normal healthy stems inside a microwave oven."[26] This is interesting because water molecules are good absorbers of microwave energy, and the water content of growth nodes is higher than in any other part of a plant. Dr. Haselhoff has also shown why balls of plasma might provide the energy for some crop circle formations by plotting growth node lengthening across circles. He found the maximum lengthening of nodes at the center, where the energy from a plasma ball would be greatest, with node lengthening trailing off toward the edges of circles. So plasma balls seem a reasonable, yet unconfirmed, explanation for simple circles. But what about those highly complex artistic patterns that appear year after year? The organization called the Circle Makers claims to have produced all of them—at least all the interesting ones. But has it? Who would finance this costly exercise, making circle after circle, year after year, and never being caught in the act? The problem is that plasma balls (and expanding fungal growth that can produce rings in vegetation) are not all that good at making highly artistic patterns. Therefore, to explain all circles you need other mechanisms: either dedicated hoaxers or . . . dare I say it . . . highly artistic aliens.

Nancy Talbot, president of BLT Research Team, maintains that ruptured growth nodes are a good indication that a circle is "genuine," although being "genuine" does not explain how it was created and who did the creating. However, the leading researchers agree that these effects are not found in circles known to be the work of hoaxers. At least that seems to be progress, but there is another complication. It isn't just the complete circles that show growth node lengthening and rupturing and soil abnormalities. Randomly downed crops can also have these abnormalities. Talbot says: "This non-geometrically downed stuff often shows exactly the same plant/soil changes documented so thoroughly in the geometric events."[27] So how can we explain these "randomly downed crops" with the characteristics of "genuine crop circles"? Are they failed attempts by artistically oriented aliens to do a good job, or are they produced by an unknown physical mechanism that does not always create circles? Or are the research finding so far misleading?

The meteoric dust found in crop circles provides another problem. A member of the BLT team, John Burke, reports that "microscopic particles of meteoric dust [were] found in two-thirds of the 32 formations where we have been able to obtain soil samples."[28] He suggests that this points to plasmas formed in electrically charged regions of the upper atmosphere. There are no definite findings yet, but even if this natural mechanism was proved to exist, the complex circles would remain unexplained.

Dr. William Levengood (the "L" in the BLT team) has studied crop circle material at his Pinelandia Biophysical Laboratory for many years and has published three peer-reviewed papers on the subject. He was the first to show that the growth nodes of plants had elongated, and in some cases ruptured, because the fluid within them had expanded on heating. Levengood's work led to the formation of the BLT Team that seems to have taken the lead in crop circle research. Their latest contribution is a comprehensive project conducted by eight researchers from Canada and the United States. Their detailed paper, published in 2004, provides results on the study of barley plants and the soil at a seven-circle formation that appeared in Edmonton, Canada, in 1999. But one study like this is not enough, so we still don't know what is going on.

The only way forward is to do more research and get repeatable results, if that is possible. If enough scientists independently discover consistencies, then we might be on our way to explaining crop circles. But those scientists shouldn't let anyone think they're investigating the handiwork of aliens—not if they wish to continue to move in scientific circles as well as crop circles.

WHO'S CARVING UP OUR CATTLE?

Another unexplained—and weird—phenomenon that some people associate with UFOs is that of animal mutilations, though it's hard to reconcile this with other aspects of the UFO phenomena. All one can say is that it at least has farming in common with crop circles. I suppose the basic question for ufologists is: How can one lot of aliens be entertaining us with beautiful circular art while another lot are mutilating our cattle? The reasonable explanation is that the cattle have died naturally and that hungry rodents and birds are doing the mutilating. It's the accessible genitals and udders and eyes that most often go missing. Vets point out that wildlife can do a neat nibbling job, while some ufologists say that only laser surgery could be responsible. However, if aliens were responsible, why do they need so many samples of cattle anatomy? The phenomenon is frequently reported, which makes sense if birds and rodents are responsible because cattle do die in their fields. Also, cattle mutilations are reported mainly in the United States where the UFO phenomena are well publicized and people are quick to blame the aliens if UFOs are in the news. Of course, this weird subject is like any other that makes colossal claims. If you put forward the hypothesis that aliens are killing cattle for their experiments, you have to provide proof to match the claim. So far this hasn't happened, although the National Institute for Discovery Science in the United States did carry out some very detailed scientific research and published the results on its website. But it hasn't provided an answer. It's difficult to see how it could.

THE SCIENTIFIC OPTIONS

Before we leave this chapter, let's briefly review the different options for searching for the signatures of life.

(a) Astronomical SETI, both radio and optical, depends on a long series of dedicated broadcasters in the galaxy, dating back to about four billion years ago when, in theory, the first technological creature could have evolved. The main advantage of this approach, with several billion sunlike stars in our galaxy, is that it has numbers on its side. But knowing the odds against success, is anyone going to broadcast? However, judging from the discoveries of planetary systems in recent years, there could be billions of planetary systems out there, and someone might be doing something that is astronomically detectable. Astronomical SETI would also be needed to detect evidence of alien probes within the solar system.

(b) The scientists in "Planetary SETI" may find evidence in close-up photographs of the Moon and Mars. Their work has identified interesting sites—though no more than "interesting"—that justify more detailed photography of those two worlds. This is a promising development because the window of opportunity for visits to the solar system is so vast. Objects left on the Moon a billion years ago might still be detectable because of the virtual absence of erosion. The checking of video evidence by independent teams of analysts (preferably in university departments) might confirm or reject the conclusion that alien artifacts are flying in our atmosphere in ways that would be impossible for our own aircraft.

(c) A fully coordinated scientific investigation of the physical traces that witnesses find at UFO landing sites might discover patterns of chemical, physical, and biological evidence, so making predictions possible of what might be found at future reported landing sites.

(d) The thorough analyses of strange artifacts, found from time to time, to see if they can be explained as belonging to our own environment. The scientists at the Research Institute on Anomalous Phenomena in Ukraine call this work paleo-visitology. They have analyzed a few interesting items but have found nothing extraterrestrial—so far. There is also the study of historical material.

(e) The science community has left the mystery of crop circles to remain a mystery. If a natural or human-made mechanism for their formation is not found, the claims that alien visitors are playing games with us will grow. The circles offer ready access for scientists to take all the samples they need for testing. Most other possible manifestations of extraterrestrial activities don't hang around to be so easily investigated.

(f) Cattle mutilation is bizarre. It apparently happens all over the world but is mainly documented in the United States. The mutilations tend to focus on specific parts of cattle anatomy, including the eyes, udders, and genitals, and evidence of surgical dexterity has been claimed. So is it due to innocent hungry birds and mice, taking advantage of an easy meal? In some cases, farmers have reported strange lights at night in fields where, the next morning, they have found dead cattle, but are these the lights of the mutilators? No long-term scientific study has been carried out, and no acceptable answers are available. Some say it's all the work of psychopathic hoaxers. Others say it's psychopathic aliens. Who would you rather have operating outside your house?

(g) The phenomenon of alien abductions, with an estimated two million people claiming to have been abducted by aliens, looks like a widespread psychological phenomenon kept going by the media. The reason: there are so many cases yet no forensic evidence. How do the aliens do it? Until forensic scientists repeatedly discover evidence on the clothing of abductees, the conclusion must be that the aliens are not guilty. What we need in all these options are both rigorous scientific research and repeatable results. They can come from a wide range of possible and improbable phenomena: alien broadcasts, unexpected spectral lines from Dyson-O'Neill spheres, the detection of alien probes in Earth or solar orbits, chemical traces at reported UFO landings, artifacts on the Moon and Mars—or even abductions, providing there is verifiable proof. Amid all the confusion and skepticism, SETI is really in a favorable position because it has such different and independent paths to follow in testing the Grand Hypothesis. And at this stage no path should have overwhelming priority over the others. Search for evidence of probes; or ancient artifacts on the Moon and Mars; or the physical, chemical, and biological analyses of reported UFO landing sites are just as worth doing as searching for hypothetical broadcasts from across the galaxy. Evidence may or may not exist, but scientific research is the way to find it if it does.

CHAPTER 7

MYTHOLOGY
AND REALITY

Taking a trip in a flying saucer looks like the ultimate in "getting away from it all." And there's the bonus that some people will believe your story. What the aliens have to say about saving the Earth from pollution, global warming, and nuclear war will keep your name in the UFO literature for years to come. You may even be invited to UFO conferences to tell your story. Most "contactees" who have done so believe their fantasies, but those awkward skeptics want evidence. Of course we can't say that no one has ever seen a UFO occupant or had contact with one. We can only say that there is as yet no acceptable evidence of this. We need more than a detailed report to accept such paradigm-shifting information. Actually, those who fantasize about contacts with aliens usually undermine their stories by piling on the details: the more elaborate contact stories become, the less they ring true. Some people are just not content with one contact; they want regular meetings with alien visitors. So their stories gain momentum in the media and blossom into permanent mythology, as the "abduction phenomenon" has done. These days—and usually under hypnosis—people report a range of uncomfortable experiences during abductions. They don't just get a ride in a flying saucer as they used to when the idea was new. The mythology has changed. Anyone taken aboard a flying saucer these days can be the subject of some nasty medical experiments. It wasn't like that years ago. At one time you could have a good time with the aliens, visit a couple of planets over the weekend, and get your name in the newspapers along with the celebrities. I'm not exaggerating. I've known this to happen. Well, the couple I knew didn't actually visit a planet, but they did talk to an alien and see his flying saucer—and get their story in the national newspapers and on BBC television. Their adventure was even quoted by an innocent Australian scientist in a serious article on the UFO phenomena. Now, some of the nicest people are fantasy prone, but their reports should not be accepted by UFO societies, which then invite them to lecture about their adventures. The media don't care if these stories stand up or not. They run them for entertainment. But the UFO societies are not supposed to be in the entertainment business, and anyone who claims to have been inside a flying saucer and talked with the aliens—especially by telepathy—should not be lecturing at UFO conferences about their fantasies.

Most conference speakers are rational people trying to find answers to a mystery. Sometimes a few scientists contribute. But some speakers, who are usually agreeably gullible, are uncritically immersed in the mythology of UFOs. For example, lecturers make claims about alien artifacts that, if true, would constitute the greatest scientific discovery of our age. To proclaim such "discoveries" to the world without irrefutable evidence is the antithesis of science. And the leading UFO associations know this. One lecturer at an international conference actually claimed to have seen over fifty UFOs and to have been in communication with a UFO pilot for the past seven years, yet in all that time he had not learned anything new or anything we could check on. So while the major UFO associations proclaim their unwavering devotion to science, they regularly have speakers at their conferences who claim without proof to have confirmed the major hypothesis in human history.

The UFO societies should note that not long ago two leading physicists claimed to have shown that energy could be obtained from cold nuclear fusion ("cold" is the key word here). This would have been a convenient departure from the present struggle to harness the same process of energy production that keeps the Sun and stars shining. Physicists who have been trying to reproduce this energy source in experimental nuclear reactors for decades must contain hydrogen plasma at fantastically high temperatures—so high that the plasma (the energy source) can't be contained for more than a fraction of a second. Cold fusion, if that were possible, would obviously have bypassed

this unsolved problem. So teams of physicists got to work and tried to duplicate the original "cold fusion" experiment. No one could make it work and
the subject was eventually written off. That was the response of the science
community to a major claim from two top physicists. Had cold fusion been
confirmed it would have changed our lives, as would confirmation of evidence
of extraterrestrial intelligence or that some UFOs have a physical reality. You
will find people in ufology who accept that a large part of their subject is
mythological while maintaining that science too is a myth created by the
scientific method. The difference, of course, is that no one has ever applied a
myth to create a new technology, whereas science is constantly used to create
new technologies. And you can test for any item of accepted scientific data if
you doubt its reliability. You can't do that with myths. In other words, science
is a presentation of reality as far as we can know it in any given age, and the
scientific method is the only way to explore the physical nature of life and the
universe.

Science can therefore put the UFO phenomena into some sort of perspective, as I've tried to do in this book. But this is not done in the media, partly
because most journalists regard UFOs as paranormal phenomena, which
includes all weird and wonderful things from ghosts to mental telepathy. One
gets the feeling that even today, in the media mind, ghosts are marginally
more respectable than UFOs. Many rational people see ghosts, and others no
less rational who have never seen a ghost believe in them. Yet despite investigators armed with the latest in ghost-hunting technology, ghosts have never
manifested themselves at the right time and place to have their credentials
checked. There are psychological reasons, though, for belief in ghosts, and
these same reasons lead some people to believe in UFOs. The big difference,
though, is that ghosts are shockingly short of a scientific rationale, whereas
SETI science provides one for at least some UFOs being alien artifacts. Also,
as we've seen, UFOs have provided data for scientific investigations, whereas
ghostly data from the spirit world is rather difficult to test. I suppose one
thing could be concluded about ghosts—apart, perhaps, from their nonex-
istence. If they could materialize, ghosts would have to be composed of very
rarefied matter to pass through walls and the like. But we can't get a testable hypothesis out of that, though scientists have tested other aspects of the

psychic realm with no success. If the psychic realm exists, it is unknown and unexplained, and we can't use this to explain the UFO phenomena. We can't use one unknown to explain another unknown.

GOOD-BYE TO CLASSIC CASES

A ufologist once told me that classic cases have been so well investigated that it's difficult for skeptics to find weaknesses that would discredit them. That sounds more like law than science, and it's science we need to escape the fruitless situation where ufologists investigate and write up story after story while armchair skeptics try to explain them as natural events or hoaxes. This writing up of classic cases from what has already been written and published many times already has been a black hole for ufology. The endless discussions about old classic cases go nowhere, no matter how trustworthy the witnesses or how many years investigators have worked on a particular case. Roswell is the best example of this, in which a saucer is supposed to have crashed in the desert in New Mexico in 1947. You could fill a library with the books and articles written about this subject, yet not one item of physical or biological evidence has ever been put before the public. It seems especially surprising— if the story were true—that no UFO investigator has managed to track down even one biologist from among the many who would have carried out research on the alien bodies. There exist only odd remarks about people who spoke with nurses who saw doctors examine alien bodies from the crashed saucer. Neither has anyone in sixty years found a fragment of the saucer for analysis, though material from it was reportedly scattered over the desert in hundreds of pieces. The best support for the Roswell story came from Jesse Marcel, with the US military, who, after his retirement as a lieutenant-colonel, gave his version of the Roswell incident. He was a major in 1947 and was sent to collect debris from the "crashed saucer." As he was too late to deliver the collected material to his base that night, he drove home, where he and his family examined the various items. When he went public years later he described these as having advanced and unusual properties, such as metal beams "three-eighths of an inch square" that were amazingly strong and light, and thin metallic sheets that could not be deformed. Unfortunately for all future Roswell investigators he didn't keep a single piece of this evidence, so there's nothing to check to see if it was high-tech human technology or bits of a flying saucer.

The Fund for UFO Research once stated that "since 1990 the organization has spent nearly $100,000 for ongoing research on the Roswell UFO crash case." It isn't that a saucer crash is impossible, if you accept that strange aerial craft are flying in our atmosphere. But investigators have had sixty years to find a piece of alien technology or a scrap of alien biology from the crash, or any information from the army of biologists who would have created a library of research papers. The greatest adventure for any biologist would be to work on such a subject. So much could be discovered about the phenomenon of life—our life as well as the alien life. But all we have are reports of what local people have said in interviews about something that happened years ago. Some of the reports may seem credible, but they can't lead anywhere unless the US government has a few dead aliens hidden away, plus a flying saucer in need of repair. And we can't expect that.

Such aspects of the UFO phenomena tell us that we have to bypass mythology and rely on science. There are three reasons for doing so:

- The relevant science enables us to critically examine the enormous amount of reported data, most of which is flawed.
- It can help us find technical ways of testing hypotheses.
- It enables us to recognize the hoaxers and people with vivid imaginations. And the procedures of analytical chemistry, physics, microscopy, cell and molecular biology are all needed to check the data included in UFO reports.

THE WRONG QUESTIONS

Interviewers in the media will often ask questions like

"Do you think that a flying saucer crashed in New Mexico in 1947 and that the United States government has its dead occupants locked away in a freezer?"

Or: "Do you think that the Belgian air force, which claimed visual and radar contact with a UFO in March 1993, has shown that at least one extraterrestrial spacecraft was flying in Belgian airspace?"

Or: "Do you think that American Air Force personnel, who were in charge of a nuclear arsenal in Britain, chased an alien artifact through the Rendlesham Forest on the last nights of December 1980?

The answer to such questions is that we cannot know unless investigations were carried out and evidence discovered. No matter how reliable a report may seem (and we might expect the US military in charge of atomic bombs in Britain to be very reliable) reports themselves can only be the beginning of scientific investigations. And there's no evidence that such investigations have taken place. The result: endless interviews with witnesses for years after the event and no scientific information to prove or disprove anything.

Associated with this problem is the production of "paper studies." Informed people sit at tables with their notes, exchanging views and arguing with other informed people. In the end they agree, or disagree, and issue a report of their views. But at best only a few people do any practical research that might provide reliable data. And this doesn't mean going out and interviewing more witnesses and collecting more anecdotes.

The fact is that the media hardly ever gets to grips with the relevant science. Where professional scientists are mentioned in UFO reports, it's usually "what scientists found" reported in a generalized way when you want to read the words of the scientists themselves. The research could then be critically scrutinized. In all scientific disciplines a reliable database exists that has withstood critical examination. Take any subject in science, and you can find references to papers written by researchers in the field you are about to enter. You can find reliable information, discover what is already known, and see what remains to be done. The literature of ufology seldom provides this facility because decades went by with hardly any scientific research. There may be no shortage of newspaper and magazine reports, but scientific research can't be based on these. Journalists report UFO stories whether they are credible or not. No one complains that they fail the reader or listener by not covering the relevant science, which could demolish most reports. Take the film of bogus alien bodies that was claimed to have been taken soon after the Roswell saucer crash. The biology on display cried out "hoax," yet the media allowed the story to con half the world. So next time there's a UFO piece on radio or television, take note of how much relevant science it contains. Usually there won't be any, but sometimes the media does try to give a balanced view. On April 3, 1997, the national news and current affairs program in Britain, "Radio 5 Live," ran a phone-in about UFOs. It had two "experts" fielding calls. A scientist from aerospace research had left his laboratory to put a major superstition to rest. The other "expert," a well-known ufologist, supported the extraterrestrial hypothesis for some UFOs. But the discussion was about "belief" and "nonbelief." No one pointed out the link between astronomical SETI that has searched for alien probes in Earth orbit and the UFO phenomena. The radio program consisted of calls from people who thought they'd seen a UFO, with the scientist offering explanations such as aircraft lights, the Moon, Venus, and balloons. In most cases he was probably right, but one could go on like that forever.

THE FANTASY SECT

When you meet people in the UFO societies you soon realize that the fantasy sect is not large. It only appears large because the most colorful stories get well publicized in the media. Nowadays, of course, there's unlimited scope for all fantasy sects because the web offers unlimited opportunities for both good and bad websites. Some stories are there just to entertain, like one I saw in the *UFO Casebook*. A pharmaceutical salesman in Spain sees a pear-shaped UFO on some ground ahead of his car. Two humanoids are working near the craft. He stops the car and walks over. Did the humanoids need help to fix their UFO? One of the humanoids declines his offer in an alien Spanish accent. Then both enter their UFO and disappear into the sky. Another tall tale comes from *Pravda*. Fishermen in western Russia had been seeing a green flying saucer. Every time it passed overhead they had an unquenchable thirst for alcohol and arrived home drunk and fishless. UFO tales are not always so innocent. Sometimes they are menacing, with reports that the Earth is now a center of operations for several alien species that have quite recently crossed the light-years to conquer Earth. However, the CIA is on to them, so that's reassuring. Investigations are continuous, and the Pentagon is kept fully informed.

Of course, evidence is quite unnecessary in the domain of true UFO mania, as the website Crowded Skies confirms. While evidence of just one alien species would be enough to win plaudits from the entire science community, this is not enough for Crowded Skies, which offers a whole taxonomy of different alien species, including the diminutive "grays" (in various shades of gray), the creepy insectivoids, and the devilish reptiloids. Adolf Hitler, I learned, had been groomed by the reptiloids for his role as Nazi dictator. But why stop at Hitler? Such an intelligent species as the reptiloids would see lots of opportunities to shatter peace on Earth. Perhaps they were behind the communist takeover in Russia and the rise of Stalin? Then I read that, having come all the way from their home planet in the Orion constellation (or was it Draco?), their sole purpose was to cause us trouble. They might even be working on us right now, making us believe all sorts of weird things about flying saucers and their humanoid occupants, like the belief that the insectivoids are getting ready to trigger all those implants in the heads of abductees, so starting programs in their brains that will make them rule the world. For a moment I thought the mind of some poor ufologist had buckled under the weight of too many UFO stories. But it's a clever hoax . . . yes? To highlight the nuttiness that pervades UFO mythology? Probably the CIA or MI5, or is it MI6, in Britain are behind it, though that thought might seem worse than being paranoid about aliens.

WELCOME TO ALL ALIENS

Even the science community, when confronted by the intriguing prospect of alien visitors, can sometimes get carried away beyond the fringe, although perhaps not so far beyond to where the reptiloids live. Anyway, I hope the reptiloids won't get on to this, but visiting aliens can now receive a warm welcome on Earth. Well, at least from the eighty scientists and academics who have joined with Dr. Allen Tough of the University of Toronto to persuade extraterrestrial visitors to contact humanity directly through the

Internet instead of wasting their time hovering around aimlessly in those flying saucers and making all sorts of patterns in cornfields. "We welcome you and seek dialogue" is the message extended to any intelligent life-form (robots included) that has reached Earth and is wondering what the hell to do next. They need wonder no longer. "We (the eighty academics) could aid your learning by organizing some sort of structured conversation between you and some particularly insightful individuals." A "structured conversation" with clever octopoids comes to mind, as do chats with the galaxy's most advanced robots. Yet Allen Tough and his associates are optimistic: "Major benefits will occur after genuine contact is confirmed, especially if some sort of dialogue occurs."[29]

We can admire such optimism. It's a big step forward when eighty prominent academics sign up to the idea of getting the aliens to contact us on the web. By comparison, simply saying that "they could be here in the solar system" looks ultra conservative.

This is old SETI chauvinism. Could we expect advanced aliens to put the *Encyclopedia Galactica* on the web for easy exploitation by such a diverse and unstable world civilization as ours? Of course not. But new technological creatures like us, as we see from the chain of fortuitous events that led to our own evolution, are going to be rare. So the aliens, if they know about us, are likely to treat us like newborn babies whose future could be interesting if left undisturbed. That's the best we can hope for—as long as the reptiloids don't get us first. However, Tough also envisages the possibility of a no-contact scenario, presumably as an alternative to "structured conversations." He writes: "the primary purpose [of visiting aliens] is to observe the natural development of our civilization, untainted by contact."[30] Judging from the elusiveness of flying saucers, this could be correct. But in this case Tough won't receive any alien e-mails, though some thirty hoaxers have tried to fool him.

His project is certainly beyond the fringe, but it does make us think about the subject. Though if those diabolical reptiloids get on the web, who knows what might happen? Better to try to detect a possible alien presence while avoiding the "missing technology" problem. In short, let's not expect "messages" that our present-day technology could receive or that we could understand. Better to consider things that shouldn't be present, such as the reported radiation from UFOs and its various effects, or physical and chemical patterns at credible UFO landing sites, or photographic evidence of strange aerial craft, or any signs of artificial structures on the Moon or Mars. And while we do all this we can still take cosmic messages and even alien e-mails, if any arrive.

ALIENS NEAR AND FAR

Of course, for aliens to send us e-mails they would have to be nearby aliens and active aliens, like the saucernauts of some UFO reports, or hibernating robots who, after millions of years, have just been switched on by some trigger in our environment. Of course, most people would not want the Grand Hypothesis confirmed by the presence of nearby aliens. Really distant aliens would be far more acceptable. The more light-years away they were, the safer they would appear. Having alien craft flying overhead, bus aircraft, and land in front of cars is rather different from having super beings broadcasting their wisdom across the galaxy for our benefit. But definite confirmation that "they" are here, or have been here in the distant past, would nevertheless be a positive result for the Grand Hypothesis. The universality of life and intelligence would no longer be a hypothesis but a fact, although not such a comfortable fact as the SETI astronomers had hoped for. Evidence of planetary civilizations thousands of light-years away would not demolish our present paradigm in so shocking a way. It would just modify it a little. But if alien artifacts are in orbit and landing on Earth in our epoch, then other intelligences (biological or robotic) have been in the solar system observing the Earth for an eternity, and the phenomena of life and intelligence would have to be very widespread. For statistical reasons it could hardly be otherwise. And we would know that we are of considerable interest to at least one far older civilization. We could also conclude, on the positive side, that we have great rarity value because the Earth has been left undisturbed. But how would governments and the population at large respond to the biggest paradigm shift in history? The general public would need to know that as the aliens have had eons to invade we should be perfectly safe. World-conquering aliens could have taken over any day during the past few hundred million years with no more opposition than dumb animals. They haven't done so and are not likely to now. Of course, the strong possibility remains that UFOs are all mythological and that there is nothing more than some new plasma physics to discover from those strange lights in the atmosphere. At least the British government's UFO Study would have us believe that this is so. But the chance of a new reality hidden by the mythology and fantasy is high enough for professional scientists to get involved—and some are getting involved for the reason we've seen earlier. What no one can deny is that the science that supports astronomical SETI, a respectable scientific discipline, equally well supports the hypothesis that some UFOs are alien artifacts.

INEXPLICABLE TECHNOLOGIES

We know that mythologies blossom from the strange and inexplicable, which is precisely how the technologies of another world would appear to us. So the astonishing behaviour of strange aerial craft should not make us disbelieve witnesses when such strangeness should actually raise our interest. Let's pursue a little time travel again and bring into our time a few ancient Greeks, learned citizens of Athens from three thousand years ago. They would see television, computers, telephones, electric lights, aircraft, and motor vehicles. They wouldn't comprehend what they saw because they would be three thousand years away from knowing the science on which these things are based. Yet we are close in time and culture to the ancient Greeks—and of the same biological species—while the cultures and technologies of other worlds would be vastly different from anything in Earth history. The SETI scientists should allow for this and be ready to investigate data we can't explain—just in case it holds the answer to our main question. We can't leave the major UFO associations to do this. They pay lip service to science but have rarely engaged in scientific research. I quote from the largest UFO association, the Mutual UFO Network (MUFON): "MUFON boldly states that a concentrated scientific study by dedicated investigators and researchers will provide the ultimate answers to the UFO enigma." I read this years ago, and the statement was still on the MUFON website when I checked recently. But ask MUFON for scientific research papers of the kind prepared by all working scientists. This would tell you that MUFON's statement is partly misleading. However, science does begin with data collection, and there are some excellent catalogs of data produced by MUFON members and those of other associations. Decades of investigations have produced valuable catalogs on every aspect of the UFO phenomena. For example, MUFON's computer database, called UFOCAT, which was begun years ago, contains more than 170,000 entries and is claimed to be the most comprehensive catalog of raw data. This is admirable, but the collection and verification of reported data, although an essential basis for research, is not scientific research in itself.

However, this good work of data gathering is undermined at major conferences by lectures on abductions, telepathic chats with the pilots of flying saucers, and secret alien bases in obscure South American locations. All this is light-years away from MUFON's "concentrated scientific study." At a MUFON conference in Denver, a report described how a large UFO flew over a part of Mexico and was captured on video and shown on TV. Local villagers eagerly confirmed this event, though a video technician had shown the tape of the UFO to be a hoax. One witness said "the cats and dogs went crazy in the presence of the UFO, and a parrot that had never talked started talking." It was also claimed that the parrot was in telepathic communication with the UFO occupants, which was not unexpected. If humans can communicate this way, as many believe, why not parrots? We have since learned that the parrot was later abducted from its perch and that no one has heard from it since—not even by mental telepathy. Needless to say, this sort of thing deters scientists who don't want to get involved with telepathic parrots.

BOOKS BEYOND THE FRINGE

This brings us to the literature of ufology where many books lack credibility. It seems that most authors are innocent enough—just high on the gullibility scale. They believe what they write. It's just unfortunate that the subject requires firm critical faculties. Yet the people who put their stories into print are not so blameless. They must have good critical faculties to survive in the media. They know nonsense when they see it. But if it's good commercial nonsense, they publish. Some years ago a major publisher brought out a sensational book on a well-publicized case: the Eduard Meier case. Meier, a Swiss caretaker, had been conversing with a fashionably dressed lady ET who regularly visited him in her flying saucer. In her photo she looked about as alien as Miss World on a day out, yet Meier claimed she came from the Pleiades star cluster. This should have been enough to discourage the most gullible investigator, as well as every publisher on the planet. Not so. A book was written, and, at its launch, dozens of copies were given to journalists who were then shown Meier's spectacular films of saucers in flight. Parts of the British national press were sufficiently bamboozled to proclaim the photographs the best evidence yet that alien spacecraft were with us. The publisher was happy. Meier was happy. The book's author was happy. The journalists were happy. But the public was taken for a ride, though not in a flying saucer.

Respected ufologists, who had nothing to do with this farce, were furious. Some months later a well-known UFO investigator wrote to me after reading my review of her latest book. She complained that ufologists like her suffered because the media constantly promoted hoaxers and psychologically generated UFO experiences. She also complained that serious UFO research received little coverage in the British national press compared with UFO nonsense. She went on to say that her research group had shown the true nature of the Meier case six years previously, when models of flying saucers (the reality behind the photographs) were discovered in a garage of one of Meier's neighbors. Actually it wasn't necessary to find those models. The scientific information that the glamour puss from the Pleiades had regularly given to Meier proclaimed a con. There was no sensible science. Apart from appeals to save the world from war and pollution and to unite humanity in peace, a school pupil well soaked in science fiction could have invented everything. Yet the book's author, being both a lawyer and an investigative journalist, didn't seem to notice anything wrong. Of course, it's not easy for nonscientists to recognize phony scientific information, but that's no excuse in this case. With the general public, it's different. Pseudo-scientific jargon is everywhere in advertising. It's nonsense. The advertisers know it's nonsense. But as long as most people don't recognize it as nonsense, it works. It sells the product.

ONE WAY FORWARD

Only one thing from the fantasy-prone history of the UFO phenomena seems clear. If we want to know the truth we have to apply science and avoid all things mystical, telepathic, and pseudo-scientific. Go down that road, and you end up talking to Meier's glamorous lady from the Pleiades and to those weird reptiloids, insectivoids, and amphiboids—and even telepathic parrots. We don't need any of that if we're going to get answers to the questions posed by the UFO phenomena. We have to keep in mind that science has paid off in the past and will do so again, if there is anything to discover. Yet many people still regress to the mystical mindset of our medieval ancestors when confronted by the unknown. This tendency, which is widespread in ufology, has damaged its standing, while the SETI astronomers have forged ahead within the science community, which is armor-plated against anything paranormal.

So can the major UFO associations put the emphasis on science and not on spectacular claims that come with no evidence to support them? Can they avoid those telepathic communications with aliens and free trips in flying saucers? Can they free ufology from the lethal grip of the insectivoids, reptiloids, and amphiboids? Can they move on from crashed flying saucers that leave not a trace behind? And can they insist that anyone who claims knowledge of dead aliens supply tissue samples or at least some believable biological data?

If we adopt a scientific approach, even the best witness reports cannot be accepted as true, although the data they contain may be useful. An official in a major UFO association once wrote to me about the absolute credibility of the abductees he'd been interviewing. He was so convinced that he accepted their stories without further evidence. It was unreasonable of me to expect more than what the abductees reported—in most cases under hypnosis. Of course, if the UFO associations insisted on a rigorous scientific approach, they'd lose many of their "true believers." But they can't claim a scientific approach unless they demand "testable" evidence. That's what all scientists aim for every day at work, and hardly any of them are attempting to prove the major hypothesis of our age, which is what anyone who claims the existence of extraterrestrial artifacts or dead aliens is attempting to do.

My guess is that the UFO associations aren't going to change, that they are not going to provide confirmation of the Grand Hypothesis. If it's ever confirmed, I think the SETI scientists will confirm it, perhaps by looking for alien probes in Earth or solar orbits, or signs of alien structures on the Moon or Mars, or searching for evidence of possible Dyson-O'Neill spheres—or even by detecting an intelligent signal from across the galaxy. These are all signatures of life in the universe. But the rational ufologists, after years of data gathering, should be given credit for their basic contribution, even though scientists with the essential know-how and tools might make the decisive discoveries. Although people love mysteries, they also want them solved, and speculations on the mystical and the paranormal have never provided answers. Ghosts and spirits are as unknown today as they ever were. But some people feel uneasy about science that might change cherished beliefs. It's true that science has made us uncertain of our position in the universe, but this uncertainty is the price to pay for current and future knowledge. And science provides our only real protection against the hazards of life, both biological and physical, something that speculations on the paranormal can never do. In our minds we were once at the center of all things, the unique creation of God. But astronomical photographs now show billions of galaxies in the universe, many of them spiral galaxies like our own, each containing billions of stars like the Sun. How many planetary systems with Earth-like planets exist in this universe of galaxies we cannot know, but the potential abodes for life seem next to limitless. And it looks from what has happened here that the universal physics and chemistry can create life where conditions allow. No wonder, therefore, that the status of life on Earth is uncertain. But a unique species evolved here a quarter of a million years ago with the capacity to find out what that status is. Perhaps not to understand life and the universe completely, we may not be biologically equipped for that, despite the enthusiasm of theoretical physicists who search for a "theory of everything." But we should be able to go some of the way, and intelligent life somewhere (biological or artificial) may understand a lot more than we do. With the size of the universe and the number of planetary systems that now seem certain to exist, we can hardly be the only searchers after truth. So while religions and philosophies have given *Homo sapiens* great importance in the universe, science has shown that we don't know if we are potential gods or insignificant newcomers to a vast and incredible empire. That's what testing the Grand Hypothesis in the search for the signatures of life is all about: trying to discover the status of our lives and our world in the immensity of the universe.

NOTES AND REFERENCES

CHAPTER 2. BIG QUESTIONS

1. The way to monitor developments in twisted-light research is to go to the website of University of Glasgow.

2. Professor Gerard O'Neill published his book *The High Frontier* in 1977, showing in convincing detail how space colonization could take place and why it could be an inevitable development for all successful technological civilizations. The need for free energy from their sun, more living space, and no pollution problems from industry would drive these civilizations into space. This may not be relevant for us now, but it is for SETI in the search for signatures of life.

CHAPTER 3. FACING THE FACTS

1. Jack Cohen and Ian Stewart, "Where Are the Dolphins?" *Nature* 409 (February 22, 2011): 1119–22.

2. Tom Durkin, "SETI Researchers Sift Interstellar Static for Signs of Life," XILINX University Program, Spring 2004, https://casper.berkeley.edu/papers/2004-03_Xilinx_SETI _article.pdf (accessed March 17, 2013).

3. All quotations from Dr. Haines in this section can be found on his website, as well as further details of his extensive research. His catalogs of data from pilots and aircrews can also be read or downloaded.

4. Roy Dutton has produced several papers on aspects of his astronautical theory. The most comprehensive, which shows how the theory emerged from the data, is "Early Work and the Derivation of the Astronautical Theory." This could be followed by "UFO Tactics: Global SAC Activity—A Study of Tactical Techniques" and "Astro-Navigational Links with Correlated Tracks—A Pilot Study." Papers can be obtained by e-mailing Mr. Dutton at roy@ roy37.orangehome.co.uk.

5. Ronald N. Bracewell, "Calling All Stars," *Science* 258, no. 5084 (November 1992): 1012–14.

6. J. Allen Hynek, letter to *Science* magazine, August 1, 1966, http://amasci.com/weird/end.html (accessed March 17, 2013).

7. "Transgenics . . . and Reverse Incrementalism," Biblioteca Pleyades, http://www.bibliotecapleyades.net/vida_alien/esp_vida_alien_18zw.htm (accessed March 17, 2013.

CHAPTER 4. THE LIFE OF ALIENS

The first four websites listed here describe various research projects to investigate the status of our genetic code and the possibility of other codes for life on other worlds.

1. After a long chat with Steven Freeland when he was in England, I have followed his website at Maryland University. This line of research is fundamental to our understanding of the phenomenon of life.

2. Professor Steven Benner, at the University of Florida, provides some excellent web pages on his group's research projects.

3. Professor Laurance Hurst is at the University of Bath, England. This paper can be downloaded from his site.

4. This paper can be downloaded from the website of the Scripps Research Institute, where Floyd Romesberg's group has been artificially mimicking aspects of the genetic code.

5. Philip J. Corso, *The Day after Roswell* (New York: Pocket Books, 1997).

6. Ibid.

7. Webb Hubbell, *Friends in High Places* (New York: W. Morrow, 1997).

8. I first encountered *Pelagibacter* when Dr. Hazel Barton did a program for the BBC. Visit her website at Northern Kentucky University for more information. The website of Professor Steven Giovannoni, at Oregon State University, also deals with related aspects of this subject.

9. Ibid.

CHAPTER 5. WHERE ARE THEY?

1. John H. Wolfe, "Complementary Document 1: Alternative Methods of Communication, in SP-419 SETI: The Search for Extraterrestrial Intelligence, NASA, http://history.nasa.gov/SP-419/s3.1.htm (accessed March 27, 2013).

2. This is a developing aspect of SETI that is worth following. Dr. Conroy's research is

available on more than one website, so if you enter his name plus "astronomy" into a search engine, you will find his information.

CHAPTER 6. TESTING TIME

1. The astronomical projects to search for images of planets and spectral lines that may indicate the presence of life are best checked on the web. Technological developments and economic factors frequently lead to projects being changed.

2. Ufology and Science, http://rr0.org/time/1/9/8/9/Swords_ScienceAndHet_JUFOS/ UfologyScience/index.html (accessed March 17, 2013).

3. The Research Institute on Anomalous Phenomena (RIAP) is, as far as I know, the only research center devoted to the study and investigation of what we can call "local SETI," including the UFO phenomena. It can call on the services of a range of academics and publishes a quarterly academic journal.

4. An online search for "Ted Phillips UFO" will give you his sites and an enormous amount of data. The essential questions are: Who studied the physical traces? How long was this done after the event? And is there any existing evidence?

5. Go to Bill Chalker's websites for an authoritative review of the UFO phenomena in Australia and the subject in general.

6. James McCampbell was probably the first scientist to show how basic physics and chemistry could explain the experiences reported by some UFO witnesses. That was about twenty-five years ago, and little has been done since then to follow up on his work.

7. "Searching beyond the Water Hole," Ask Dr. SETI, SETI League, http://www .setileague.org/askdr/waterhol.htm (accessed March 17, 2013).

8. The RB-47 event has websites where the original reports can be studied. Just search online for "RB-47 UFO."

9. Robert Sanders, "Stars Rich in Heavy Metals Tend to Harbor Planets, Astronomers Report," UCBerkeleyNews, July 7, 2003, http://www.berkeley.edu/news/media/releases/2003/07/21 _stars.shtml (accessed March 17, 2013).

10. Dan Werthimer, quoted in John Marshall, "Aliens Online," http://crash.ihug. co.nz/~marshall/seti.htm (accessed March 29, 2013).

11. Dan Werthimer, quoted in Tom Durkin, "SETI Researchers Sift Interstellar Static for Signs of Life," available at https://casper.berkeley.edu/papers/2004-03_Xilinx_SETI_article .pdf (accessed March 29, 2013).

12. Monte Ross, "The New Search for E.T.," *IEEE Spectrum*, November 2006, http://spectrum .ieee.org/aerospace/robotic-exploration/the-new-search-for-et/0 (accessed March 17, 2013).

13. Mark J. Carlotto, "The Martian Enigmas," Mark Carlotto Project Home Page, http:// carlotto.us/martianenigmas/index.shtml (accessed March 17, 2013).

14. "UFO Research Is Serious Science in the Ukraine," Anomalies.Net, http://www .anomalies.net/archive/cni-news/CNI.0413.html (accessed March 17, 2013).

15. Alexey V. Arkhipov and Francis G. Graham, "Lunar SETI: A Justification, in *The Search for Extraterrestrial Intelligence (SETI) in the Optical Spectrum II*, edited by Stuart A. Kingsley and Guillermo A. Lemarchand, *Proceeding of SPIE* (June 1996).

16. Alexey V. Arkhipov, "Archaeological Reconnaissance of the Moon: Results of SAAM Project," Society for Planetary SETI Research, http://spsr.utsi.edu/articles/Saam/SAAM.HTM (accessed March 17, 2013).

17. A. V. Arkhipov, "Archaeological Reconnaissance of the Moon: Results of SAAM Project," Astronet, http://www.astronet.ru/db/msg/1177539/e-index.html (accessed March 17, 2013).

18. Ibid.

19. Readers interested in these strange lights should download Dr. Teodorani's research papers at http://www.scientificexploration.org/journal/jse_18_2_teodorani.pdf and http:// www.itacomm.net/PH/2009_Teodorani.pdf (accessed March 17, 2013).

20. Ibid.

21. Ibid.

22. David J. Turner, "The Missing Science of Ball Lightning," *Journal of Scientific Exploration* 17, no. 3 (2003): 403–496, http://www.scientificexploration.org/journal/jse_17_3_turner.pdf (accessed March 17, 2013).

23. Teodorani research papers.

24. Ibid.

25. "Derrel Sims and Implants," PrimoContatto.net, http://www.primocontatto.net/ articles/simsimplants.html (accessed March 17, 2013).

26. "Balls of Light . . . Created Crop Circle," Biblioteca Pleyades, http://www.biblioteca pleyades.net/circulos_cultivos/esp_circuloscultivos21.htm (accessed March 17, 2013).

27. "The Latest from BLT Research's Nancy Talbot on 2003 Crop Circles," Crop Circles, http://www.jerrypippin.com/crop_circles.htm (accessed March 17, 2013).

28. John Burke, "The Physics of Crop Formations," 1998, http://www.bltresearch.com/ published/physics.html (accessed March 17, 2013).

29. Allen Tough, "How to Achieve Contact: Five Promising Strategies," http://ieti.org/ tough/articles/strategy.htm (accessed March 17, 2013).

30. Ibid.

BIBLIOGRAPHY

As far as I know, this is the only science-based SETI book that shows why there may be no messages that we can detect from other worlds and why the search for evidence of alien artifacts in the solar system may possibly succeed. To reach this conclusion we only have to consider established science and the way technology advances, which is what I have done. However, there are some good SETI books, different from mine, in which this line of thinking is not pursued, although they do provide excellent introductions to the relevant science. They include

Evolving the Alien, Jack Cohen and Ian Stewart (Ebury Press, 2002).
Just Six Numbers: The Deep Forces That Shape the Universe, Martin Rees. Britain's Astronomer
    Royal gives a clear and authoritative account of the fundamental constants.
Life on Other Worlds and How to Find It, Stuart Clark (Springer, 2000).
Rare Earth, Peter D. Ward and Donald Brownlee (Springer, 2000).
Sharing the Universe, Seth Shostak (Berkeley Hills Books, 1998).

RECOMMENDED
WEBSITES

The websites of research institutes and associations listed here are well worth exploring. Personally, I use names rather than the addresses of specific web pages, which are often long and easily mistyped. Once on a site one can usually find what one needs. But readers are welcome to contact me at EdwardAshpole@aol.com about any topic in this book.

All-Sky Optical SETI
Center for UFO Studies
Columbus Optical SETI Observatory (COSETI Observatory)
Harvard SETI
Earthlights Association International
Interplanetary Society
NASA Astrobiology
Open SETI
Optical SETI at Berkeley
Planetary Society
Project BAMBI: Amateur SETI
SETI at Home
SETI Australia Centre
SETI Institute
SETI League
Society for Planetary SETI Research
Space Studies Institute
UFO Evidence
UFO Skeptic
Worlds of David Darling

576.839 ASHPOLE
Ashpole, Edward.
Signatures of life :
R2000981619 PALMETTO

ODC

Atlanta-Fulton Public Library